Surbhi Sharma

**Membranes for Low Temperature Fuel Cells**

## Also of interest

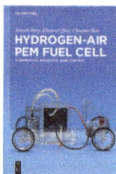

*Hydrogen-Air PEM Fuel Cell.*
*Integration, Modeling, and Control*
Tong, Qian, Huo, 2018
ISBN 978-3-11-060113-8, e-ISBN 978-3-11-060215-9

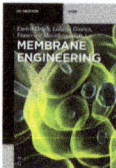

*Membrane Engineering.*
Giorno, Drioli, Macedonio, 2018
ISBN 978-3-11-028140-8, e-ISBN 978-3-11-028139-2

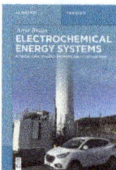

*Electrochemical Energy Systems.*
*Foundations, Energy Storage and Conversion*
Braun, 2018
ISBN 978-3-11-056182-1, e-ISBN 978-3-11-056183-8

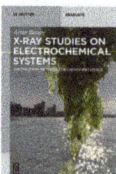

*X-ray Studies on Electrochemical Systems.*
*Synchrotron Methods for Energy Materials*
Braun, 2017
ISBN 978-3-11-043750-8, e-ISBN 978-3-11-042788-2

Surbhi Sharma

# Membranes for Low Temperature Fuel Cells

New Concepts, Single-Cell Studies and Applications

2nd Edition

**DE GRUYTER**

**Author**
Dr. Surbhi Sharma
University of Birmingham
School of Biosciences
Edgbaston B15 2TT
United Kingdom
s.sharma.1@bham.ac.uk

ISBN 978-3-11-064731-0
e-ISBN (PDF) 978-3-11-064732-7
e-ISBN (EPUB) 978-3-11-064739-6

**Library of Congress Control Number: 2019934275**

**Bibliographic information published by the Deutsche Nationalbibliothek**
The Deutsche Nationalbibliothek lists this publication in the Deutsche Nationalbibliografie;
detailed bibliographic data are available on the Internet at http://dnb.dnb.de.

© 2019 Walter de Gruyter GmbH, Berlin/Boston
Typesetting: Integra Software Services Pvt. Ltd.
Printing and binding: CPI books GmbH, Leck
Cover image: designsstock / iStock / Getty Images Plus

www.degruyter.com

# Preface

Proton-conducting membranes (mostly polymer based) are responsible for separating the anode and cathode reactions in low-temperature fuel cells and, thus, lie at the heart of a fuel-cell system. These membranes must also possess attributes such as long-term tolerance towards chemical and thermo-mechanical degradation in harsh fuel-cell environments involving temperatures from room temperature to >100 °C, variable humidity as well as variable pressure conditions. The stability of these membranes over several hundred to thousands of hours is essential for achieving consistent and optimal fuel-cell performance. Moreover, the crossover of reaction gases (hydrogen and oxygen) must also be minimised to attain long-term stability and prevent the formation of 'hotspots' occurring due to the reactions between gases in the presence of migrated electrocatalyst particles, which can lead to 'pinhole' creation and trigger membrane degradation.

Consequently, over the last few decades, polymer- and non-polymer-based materials have been researched extensively for use in membranes for low-temperature fuel cells. Nafion®, a type of perfluorosulfonic acid (PFSA) membrane among other commercial PFSA products, forms the current commercial standard. There is, however, substantial scope for improvement given the demands of the fuel-cell industry in terms of operability and durability at a wide temperature and humidity range for thousands of hours of operation. A myriad of fluorinated and non-fluorinated polymers as well as hydrocarbons and natural polymers have been explored specifically for proton-exchange membrane fuel cells (PEMFC). Composite membranes using polymer as well as non-polymeric materials (graphene, clays, solid acids, ionic liquids (IL), polymeric IL, nanostructured silica, zirconia, ceria) as fillers have been studied widely. They achieve and augment different features and characteristics, such as water adsorption and retention for low-humidity and high-temperature operation; range of operating temperatures; suitability for sub-zero operation; mechanical strength; gas permeability and methanol selectivity to reduce crossover.

In addition to composites, various novel membrane concepts such as multilayer membranes, direct membrane deposition, and electrospinning have also garnered attention in PEMFC over the last few years. These concepts have been materialised using a range of preparation methods, including dip coating, hot pressing, solution casting, and electrospinning.

Testing of these novel membranes using several physical characterisation, durability and degradation tests outside (*ex situ*) and inside (*in situ*) an operating fuel cell is of utmost importance. *Ex situ* studies are usually carried out using physical characterisation methods such as X-ray photoemission, electron microscopy, X-ray diffraction, thermogravimetric analysis, differential scanning calorimetry, nuclear magnetic resonance, and vibrational (Raman and infrared) spectroscopies for standard polymer and material characterisation. Other *ex situ* tests such as tensile testing, determination of water uptake and ion-exchange capacity, gas permeability,

https://doi.org/10.1515/9783110647327-201

and proton conductivity are also undertaken regularly to understand membrane behaviour and properties. These *ex situ* tests provide valuable information on novel membranes before proceeding to *in situ* studies, thereby saving precious time and resources.

*In situ* studies using single and multi-cell stacks involving current voltage and power curves as well as in-depth studies involving humidity and temperature variations (relative humidity cycling, freeze/thaw testing) and examining chemical and mechanical durability by artificially introducing chemical contaminants and physical defects have been reported extensively for Nafion® and other PFSA membranes. These tests examine chemical, thermal and mechanical stability. However, they can also be combined with various physical characterisation and simulation methods to collect data and visualise membrane behaviour and the changes therein to water adsorption, transportation and proton conduction with respect to various operational conditions and applications. These experimental methods have led to significant advancement in our understanding of membranes in the complex fuel-cell environment.

*In situ* studies reported on other membranes are limited. However, studies involving single-cell tests on different non-PFSA and non- polymeric membranes have picked-up pace in the last few years, enhancing our understanding of the scope and shortcomings of these membranes.

This book seeks to provide the latest academic and technical developments in PEMFC membranes. Chapter 1 gives a brief historical perspective of PEMFC and polymer membranes along with a basic introduction to the topics. A thorough insight into the various types of membrane materials and the array of preparation methods, along with the different characterisation and testing methods utilised, are discussed in Chapters 2, 3, and 4, respectively. Factors affecting the proton conduction, water adsorption and transportation behaviour of membranes are also deliberated upon. Recent literature on *ex situ* and *in situ* studies, single-cell and stack tests, durability (chemical, thermomechanical) and membrane degradation are examined in Chapters 4 and 5. The latter also looks at special cases reported about membrane requirements under specific fuel-cell applications such as aircraft and transit-bus usage. The future potential for various trends and the possibility of recycling membranes is also looked into in Chapter 6.

This book looks at the membrane requirements and considerations from a fuel-cell perspective. As an integral part of the PEMFC, it is essential to develop an understanding of the interactions of the membrane with other PEMFC components, along with its behaviour in different fuel-cell conditions.

It is hoped that this book will provide an overview of the latest materials and membranes explored in the literature. It should further help the reader gain an understanding of the numerous factors that govern membrane behaviour and properties, and how the different membrane properties evolve under the different operating conditions and special conditions when looking at specific applications.

This book will be an interesting read for anyone (students, academics or people from industry) interested in new materials for PEMFC membranes and those concerned with the development, optimisation and testing of such membranes.

# Acknowledgements

This book would not be possible without the continuous patience, support and blessings from my husband, parents and family who have always believed in me. It is my pleasure to be able to thank them on this occasion.

A special 'thank you' to my husband, Dr. Navneet Soin, for never complaining about losing some of our time together on the many weekends that I spent working on this book and contributing more than his share of the household chores to give me more time to carry on with my work. His wonderful words of encouragement have always boosted my spirits.

I thank my father who is forever interested in the projects I take up, never being bored with the extensive details of my time and planning on all big and small things I discuss with him.

I am also very thankful to mother who, despite not being familiar with my field of work, has always shown interest in the progress being made and has forever been patient with my mood variations resulting from a bad day at work.

I am also appreciative of my brother and all my extended family for being so understanding when I couldn't call them frequently enough.

I would also like to thank Dr. Carolina Muse-Branco, who managed to contribute to one of the chapters despite her many other commitments.

Finally, I am grateful to the publishers, Smithers Rapra, for giving me the opportunity to write this book and assisting in its editing and proofreading.

https://doi.org/10.1515/9783110647327-202

# Contents

# 1 Introduction to fuel cells and their membranes

## 1.1 Introduction to fuel cells: Brief history and basics

In the wake of the dwindling fossil-fuel resources, rising air pollution in cities and the urgent need for redeeming the environment from climate change, the development and commercialisation of economically viable, alternative power generation and storage systems such as fuel cells, batteries, solar cells is imperative. Among all other alternative/green-energy generation and storage systems, fuel cells have generated considerable interest in recent decades due to their high efficiency, low temperature and ability for quick start-up, especially in transportation applications.

A fuel cell is a simple electrochemical device that utilises the chemical energy present in hydrogen and oxygen to produce electricity in the form of direct current along with water and heat as the byproducts. More specifically, hydrogen is fed into the anode, where it is dissociated into protons and electrons with the help of a catalyst. The electrons provide the electrical current as they pass through the external circuit and reach the cathode. The protons pass through the proton-conducting membrane and crossover into the cathode to recombine with the electrons as well as the oxygen (which is fed into the cathode) to generate water. Figure 1.1 is a schematic diagram of a typical fuel cell system. The discovery of the principle of fuel cell and the earliest experiments, which established the same, are accredited to the British physicist and lawyer, Sir William R. Grove (1811–1896) and German scientist, Christian Friedrich Schoenbein (1799–1868). These two scientists carried out their studies between 1839 and 1842 and confirmed that current could be produced by combining hydrogen and oxygen. Finally, in 1844–1845, Grove reported the first fuel cell-power generator, which consisted of 10 cells connected in series. The system was supplied with hydrogen generated from the corrosion of zinc in acid [1–3]. Research was continued by researchers such as Ludwig Mond, Friedrich Wilhelm Ostwald, William W. Jacques, and Emil Baur. Interest from the industrial and wider community in the concept of fuel cells developed when Francis T. Bacon developed a 6-kW fuel cell in 1959 [4–10]. The interest in this new technology was promoted generously with the United States (US) space programme (the first practical application of fuel cells) where fuel cells were used instead of the bulky, conventional battery systems. The polymer membrane fuel cells developed by General Electric (GE) were used in the Gemini and Apollo missions for producing electricity for life support, guidance and communication systems [10, 11].

Despite the success of these missions, the research and interest in fuel cells remained mainly limited to space applications only (apart from some work carried out by companies (e.g., Allis-Chalmers and Texas Instruments) and researchers (e.g., G.H.J. Broers and J.A.A. Ketelaar, G.V. Elmore and H.A.

https://doi.org/10.1515/9783110647327-001

(a)

(b)

Proton-conducting
membrane

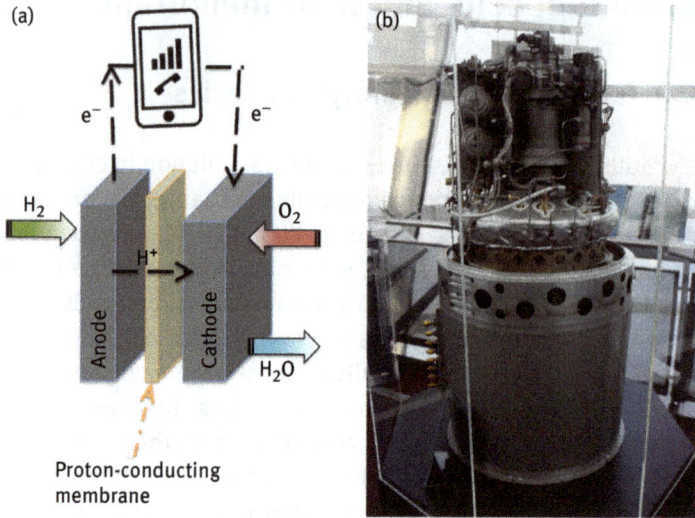

**Figure 1.1:** (a) Simple schematic showing a typical fuel cell operation and (b) a fuel cell, as used in the Apollo Spacecraft Service Module, seen at the National Space Centre, Leicester, UK. (b) Reproduced with permission from J. Humphreys [12].

Tanner, and J. Weissbart and R. Ruka in the late-1900s). It was not until the 1990s when companies such as Ballard Power Systems and Plug Power explored and demonstrated the terrestrial applications of fuel cells such as buses and cars [10, 11]. This led to the birth of an entirely new industry. Since then, with the ever-increasing necessity for alternative/green-energy sources aimed at reduction in greenhouse gases and carbon footprints, significant interest has been generated in this field over the last three decades. Today, fuel cells have found application in all types of transportation, portable and stationary devices, and power-generation systems (Figure 1.2).

In general, like any other electrochemical system, all fuel cells consist of negative (anode) and positive (cathode) electrodes and an electrolyte, which separates the electrodes. The energy-generating reactions at the anode and cathode take place at the phase boundary of the electrode–electrolyte interface, and electrons and ion transport are separated. While this setup appears very similar to that of a battery, the basic differences between the two are:

– Unlike a battery, which is a closed system in which electrodes participate actively in the electrochemical redox reactions and are consumed (marking the end of battery life), a fuel cell is an open system in which the electrodes act only as active catalysts and, as long as the gases at the anode and cathode reactions (which are consumed in the redox processes) are in supply, the fuel cell will continue to operate.

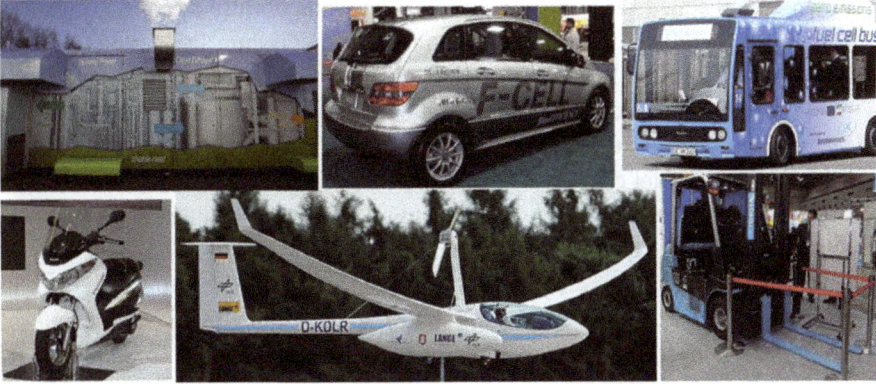

**Figure 1.2:** Commercial applications of polymer electrolyte membrane (PEM) fuel cells, also referred to as 'proton-exchange membrane fuel cells' (PEMFC'), starting clockwise from top left: stationary power generator by UTC (Reproduced with permission from WillidaUTC) [13]; car by Mercedes (reproduced with permission from Mariordo Mario Roberto Duran Ortiz) [14]; bus (reproduced with permission from Hydrogen 99) [15]; tow truck by Toyota (reproduced with permission from Comyu) [16]; aircraft (reproduced with permission from DLR German Aerospace Center in *Das von Brennstoffzellen angetrieben Flugzeug Antares/The fuel-cell powered aircraft Antares*) [17]; and two-wheeler/scooter by Intelligent Energy (reproduced with permission from TTNIS) [18].

– Batteries are discharged and rechargeable batteries can undergo several continuous charge–discharge cycles (with the help of an external electricity supply). Fuel cells are energy-generating devices that do not store charge but instead produce electricity constantly based on the continuous supply of the reactant gases which could, in turn, be stored in a battery.

Based on this simple concept and as a result of the efforts of the researchers in the 20th century, five types of fuel cell systems have emerged and which continue to be studied. These are named and classified on the basis of their electrolyte, as explained below:

– Proton-exchange membrane fuel cells (also referred to as 'PEMFC' or 'polymer electrolyte fuel cells'). PEMFC operate at 70–80 °C and use a thin, semi-porous, proton-conducting polymer membrane, which is usually made of a perfluorosulfonated acid polymer. These have found application in various portable, stationary devices as well as automotive and transport sectors. Platinum (Pt) and Pt-based alloys are commonly used electrocatalysts. For automotive applications, operating temperatures of 120–150 °C allow better efficiency. For this reason, recent research on PEMFC has attracted attention to a slowly emerging sub-category of fuel cells known as 'intermediate temperature proton-exchange membrane fuel cells' (IT-PEMFC). Direct methanol fuel cells (DMFC), which have the same setup but use methanol as the source of hydrogen, also belong to this category.

- Alkaline fuel cells (AFC) offer a wide operating temperature range of 65–250 °C which is dependent upon the concentration of potassium hydroxide (electrolyte). The electrolyte is a liquid and is retained in a matrix. AFC use various noble metals, transition metals and metal oxides as electrocatalysts. AFC were first used in US and Russian space missions.
- Phosphoric acid fuel cells (PAFC) use concentrated phosphoric acid as the electrolyte and can operate at 150–220 °C. Pt is the most commonly used electrocatalyst for the anode as well as the cathode.
- Molten carbonate fuel cells operate at high temperatures (600–700 °C). The electrolyte is composed of alkali carbonates and is retained in a ceramic matrix (α-LiAlO2). They are used for stationary power generation and combined heat-and-power systems.
- Solid oxide fuel cells (SOFC) operate at 650–1,000 °C. These systems have a solid electrolyte and have found application in stationary combined heat-and-power systems. Low-temperature SOFC, operating below 600 °C, use oxygen ion conductors such as samaria-doped ceria.

In general, all fuel cells provide 50–60% efficiency.

### 1.1.1 Polymer electrolyte membrane fuel cells: Workings and components

PEMFC (including DMFC) offer advantages such as quick start-up, portability and simplicity, which make them more popular than other types of fuel cell. Consequently, they have found utility in a large variety of applications, including micro- and portable fuel cells, for use in electronic devices (smartphones, laptops, tablets) as replacements of batteries, and automotive applications (hybrid cars, buses).

The PEM lies at the core of a PEMFC. It is a semi-permeable membrane impermeable to reaction gases and only permeable to protons (or hydrogen ions) generated at the anode. On either side of the PEM are the anode and cathode electrodes. The membrane and the electrodes together form a membrane electrode assembly (MEA). The electrode consists of a carbon-fibre cloth, which is referred to as the 'gas diffusion layer' (GDL) with nanostructured Pt or Pt alloy-based electrocatalyst ink painted on it. As hydrogen enters the electrodes through the other (unpainted) side of the GDL, it reaches the electrocatalyst lying between the interface with the electrode and the membrane. At this triple-phase boundary, the hydrogen atom is split into one proton and one electron (its primary constituents). The PEM is electronically insulating, which causes the electrons to follow an external circuit generating the desired electricity. The protons find a path through the PEM to reach the cathode side of the cell. Electrons reach here through the external circuit. Oxygen, which is supplied at the cathode, enters through the GDL and reaches the electrode–membrane interface. Just as at the anode, at the triple-phase boundary at the cathode, the three elements

(protons, electrons and oxygen) interact electrochemically to produce water. In a real system, oxygen must also be split into ions before the electrochemical reactions can take place. The basic reactions are given as:
At the anode:

$$H_2 \rightarrow 2H^+ + 2e^- \tag{1.1}$$

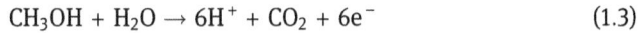

At the cathode:

$$1/2O_2 + 2H^+ + 2e^- \rightarrow H_2O \tag{1.2}$$

On the outer side of the two GDL is placed a bipolar plate (BPP), which is also called the 'end plate' or 'flow field plate'. This plate is responsible for uniform gas distribution and channelling the excess water generated at the cathode to exit the cell.

In case of a DMFC, methanol is supplied (diluted in water) at the anode instead of hydrogen but the rest of the process is similar. The basic reactions in a DMFC are:
At the anode:

$$CH_3OH + H_2O \rightarrow 6H^+ + CO_2 + 6e^- \tag{1.3}$$

At the cathode:

$$3/2O_2 + 6H^+ + 6e^- \rightarrow 3H_2O \tag{1.4}$$

### 1.1.2 Components of polymer electrolyte membrane fuel cells

The major components of a PEMFC and their primary responsibilities and functions are shown below:

- Electrocatalyst – The function of the electrocatalyst is to split the reaction gases. The efficiency of the catalyst can be affected by various factors, including the uniformity of the dispersion on the electrode, as well as its size, shape, and agglomeration. The purity of the hydrogen gas is also a major concern. Impurities such as carbon monoxide (CO) and sulfur oxides can often block the catalytic sites on the electrocatalyst, negatively affecting its efficiency over time. DMFC employ Pt–ruthenium alloys as anode electrocatalysts to minimise the poisoning effect of CO because CO can be generated as a byproduct of incomplete oxidation of methanol.
- GDL – The porous GDL plays an important part in uniformly distributing the reaction gases across the electrodes and eventually to the electrocatalyst. The material should also be an electrical conductor, to provide a connection between the electrode and BPP for the movement of the electrons/current generated. It should also be able to channel the excess water (generated at the cathode) to the BPP to avoid flooding.

- PEM – The main responsibility of this component is to separate the two halves of the electrochemical cell. An 'ideal' PEM should remain hydrated for longer to maintain appropriate proton transport. It also needs to be impermeable to gases but, in practice, membranes are not completely impermeable and, as such, 'gas crossover' is one of the issues faced by PEMFC. This problem is significantly enhanced in DMFC, where methanol crossover poisons the electrocatalysts.
- BPP – These usually have channels to handle gas and liquid water distribution and elimination from the fuel cell. BPP also connect the cathode and anode of adjacent cells in a PEMFC stack and, therefore, must be highly conductive. MEA do not have a solid, rigid structure, so BPP provide that rigidity to the fuel cell assembly. They must also be resistant to corrosion to withstand the high temperature, reactive gases and humidity conditions they are subjected to in the PEMFC. Previously, graphite or graphite polymer composites were used for BPP but, more recently, metal-based (relatively lighter and thinner than graphite) BPP are replacing them. Metal-based BPP, however, are more prone to corrosion and require protective coatings.

## 1.2 Polymer electrolyte membranes

### 1.2.1 Scope and role in polymer electrolyte membrane fuel cells

The movement of charge and its transport forms a crucial aspect of the electrochemical energy-generation process. Movement of charge from the 'source electrode', (where charged species are generated or produced) to the 'consumer electrode' (where the charged species is eventually consumed or neutralised with other species) completes the electronic circuit. As noted in the previous sections, electrons and hydrogen ions are the two charged species in the fuel cell system. Both of these species must be transported for the reactions to be completed.

Electrons are transported through the external circuit facilitated by the use of electronically conducting components such as conducting GDL and BPP along with low-resistance external circuit wiring, but ion conduction is more complex and far more difficult. This is not only due to the large difference between the mass of an electron and that of a hydrogen ion, but also the fundamentally different mechanisms of transport for the two species.

Keeping the emphasis on membranes and ionic transport as the focus of this book, further discussions will be limited to the transport and conduction of ionic species only.

Within the fuel cell system, three main forces drive the charge transport process:
- Electrical potential gradient or the electrical driving force,
- Chemical potential gradient or the chemical driving force, and
- Pressure gradient or the mechanical driving force.

Potential or voltage gradients mainly drive the electron transport in metal electrodes, but electrical and chemical driving forces have crucial roles in ion transport. Mechanical driving forces also come into play as liquid- and vapour-phase water is generated within the system over time. How these three driving forces work together is discussed in more detail in Chapters 2 and 5.

As mentioned above, apart from proton conduction, the other major function of PEM is the isolation of the two half-cell reactions by restricting the gases to their respective sides of the fuel cell. However, all current membranes offer some degree of permeability to the hydrogen gas molecules. The small size of the hydrogen molecule makes it easier for it to permeate through the PEM to the cathode side. However, significant advances have been reported in minimising the methanol crossover by increasing the tortuosity of the membrane by addition of fillers. Thus, PEM have a vital part to play within the fuel cell. Given the complex chemical environment inside the PEMFC, an ideal PEM should have at least the following attributes:
- High ionic conductivity.
- High electronic resistance.
- Chemical stability in oxidising (anode) as well as reducing (cathode) environments.
- Good mechanical strength.
- Minimal crossover of gas and fuel.
- Low cost and ease of manufacturing.

Proton conduction in PEM is highly dependent on their level of hydration due to the involvement of hydronium ions for the transport of protons (H+ ions). Hence, the ability to hold or take-up water and the temperature and time span for which the PEM can retain water are very important. Nafion$^{®}$, for example, works best under well-humidified conditions at 80 °C within PEMFC (and DMFC) environments. Humidification, to ensure the good functioning of the PEM, is usually provided by supplying humidified gases (hydrogen and oxygen) to the PEMFC system. However, the need and use of humidifiers further adds bulk and cost to PEMFC systems. Good mechanical strength is very important for the performance of PEM because they face constant cycles of expansion and compression due to the variable amounts of water at different stages of operation within PEMFC. Moreover, being the only separator component between the anode and cathode environments, it is imperative for the PEM to withstand reducing and oxidising environments. Failure of PEM was one of the major reasons of cell failure during the early years of PEMFC research, especially before the advent of perfluorosulfonic acid (PFSA) membranes.

## 1.2.2 Brief historical overview and state-of-the-art

Polymer membranes have been used for a wide variety of applications since the early 20th century. More specifically, ion-exchange membranes have found use in several industrial processes, such as: electrodialysis (desalination of saline water, the electrodialytic concentration of seawater to produce edible salt, separation of ionic materials from non-ionic materials); separators for electrolysis; recovery of acid and alkali from waste acid and alkali solution by diffusion dialysis; dehydration of water-miscible organic solvent by pervaporation [19] Based on these existing technologies, the first proton-conducting membranes for fuel cells were prepared by GE and used in the Gemini space program in the 1960s. These fuel cells (achieving 0.4–0.6 kWm-2) used the sulfonated polystyrene–divinylbenzene membranes, which were expensive and prone to rapid oxidative degradation resulting in the short lifetime of the fuel cell [20–22]. GE continued to improve these membranes but DuPont developed Nafion® in 1968 [23]. The PFSA membrane with a Teflon™-like backbone provided the much-needed mechanical and chemical stability. Following the development of Nafion®, polymer-based electrolytes also attracted tremendous scientific interest for other electrochemical applications, such as secondary batteries, sensors, actuators, supercapacitors, electrochromic displays, and dye-sensitised solar cells.

Nafion® is a copolymer manufactured by the copolymerisation of tetrafluoroethylene (TFE) and perfluorinated vinyl ether sulfonyl fluoride as well as further hydrolysis of the sulfonyl fluoride groups. DuPont also developed a general method to synthesise perfluorinated vinyl ethers from perfluorinated acid fluorides, which resulted in vinyl ether monomers. This is the starting point for Nafion® [specifically, in the case of Nafion®, perfluorosulfonyl fluoride ethyl propyl vinyl ether (PSEPVE)] as well as other novel synthesis of commercial polymers. Banerjee and Curtin provide interesting details, methods and motivation behind the success of Nafion® [24]. Today Nafion®-extruded membranes are available in various thicknesses, such as Nafion® 117 and Nafion® NR-212. The first two digits after the letter 'N' represent the equivalent weight (EW) of the membrane divided by 100, and the last 1–2 digits represent the membrane thickness in mills (1 mil = 1/1,000 inch = 0.0254 mm). EW is the weight of the polymer required to provide 1 mole of exchangeable protons, and is inversely proportional to the ion-exchange capacity (ion-exchange capacity is discussed in detail in Chapters 2 and 4). In other words, it is a measure of the ionic concentration within the ionomer. The EW in terms of g eq$^{-1}$ of a polymer membrane is expressed as:

$$EW = 100n + 446 \tag{1.5}$$

Where 'n' is the average number of TFE groups per PSEPVE monomer [11, 24, 25].

Some other companies also developed Nafion®-like membranes, namely, Flemion™ by Asahi Glass Company, Asiplex™ by Asahi Chemical Industry, Gore-Select® by W.L. Gore and Associates, Inc., Neosepta-F® by Tokuyama, and Aquivion®

by Solvay and Dyneon™ by 3M. The slight difference in these membranes comes from the number of repeating units of various monomers in the final copolymer structure, as seen in Figure 1.3. While the Teflon™ backbone provides mechanical strength to the structure, it is the $SO_3^-H^+$ group that provides the charged sites that interact with the protons and water molecules to enable proton transport.

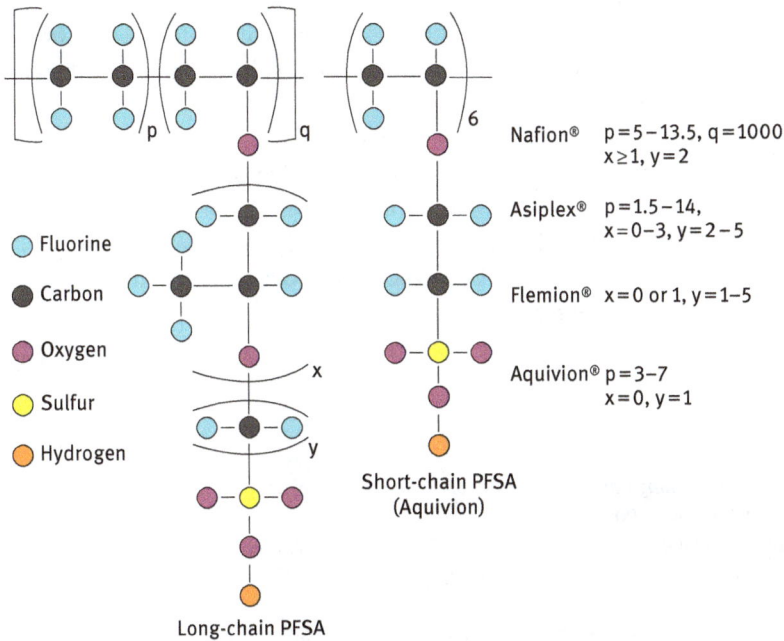

Figure 1.3: Structure of typical long- and short-chain PFSA polymers (schematic).

Over the last few decades, many types of polymer membranes have been studied extensively for use in PEMFC and DMFC. Many membrane concepts have also emerged for composite membranes consisting of various polymers (partially fluorinated membranes, non- fluorinated hydrocarbon membranes, acid–base blend membranes, aromatic membranes) as well as non-polymer, inorganic-filler materials (e.g., silica, zirconia, carbon nanotubes, graphene oxide). Layered PEM structures consisting of more than one type of polymer have also come to the fore in recent years. Nafion® is still considered the industry standard, but novel and innovative research concepts that uniquely combine new and existing materials seem to offer promising benefits for use in PEMFC, IT-PEMFC and DMFC. Despite the significant research and development of PEM, some issues that remain (or continue to persist) include: i) minimising crossover of methanol and gas; ii) increasing durability by further enhancing chemical and mechanical stability; iii) the need for self-hydrating membranes or PEM that require little hydration to reduce the need for humidification; and iv) membranes for IT-PEMFC operations suitable for temperatures ≤120 °C with minimal humidification. PEM, especially in PEMFC, have limited thickness to achieve enhanced proton conduction. Thinner membranes, however, provide limited protection to gas crossover. Moreover, thinner membranes offer less chemical stability. Membranes are under constant chemical attack within PEMFC (and DMFC) environments and, after several hundred hours of operation, metal nanoparticles from the electrocatalyst layer can be lost due to degradation of the support. These 'stray' electrocatalyst particles can attach/enter into the membrane and also react with gas molecules crossing over from either side of the cell, thereby leading to membrane degradation [24, 26].

This book discusses the various types of membranes along with the different preparation, characterisation, and *ex situ* testing methods used for membranes intended for use in PEMFC and DMFC applications. Later chapters will also delve briefly into the polymer-synthesis techniques commonly used for preparation of these PEM. We will further examine the latest *in situ*, degradation and long-term performances of existing membranes as reported in the literature.

# References

[1]    U. Bossel in *Proceedings of the 4th European Solid Oxide Fuel Cell Forum*, 10–14th July, Lucerne, Switzerland, 2000.

[2]    W.R. Grove, *Philosophical Magazine Series 3*, 1839, **14**, 86, 127.

[3]    W.R. Grove, *Philosophical Magazine Series 3*, 1842, **21**, 140, 417.

[4]    F.T. Bacon, *New Scientist*, 1959, **6**, 271.

[5]    K.R. Williams, *Biographical Memoirs of Fellows of the Royal Society*, 1994, **39**, 2.

[6]    P.B.L. Chaurasia, Y. Ando and T. Tanaka, *Energy Conversion and Management*, 2003, **44**, 4, 611.

[7]    A. Boudghene Stambouli and E. Traversa, *Renewable and Sustainable Energy Reviews*, 2002, **6** 5, 433.

[8]    J. Appleby, *Journal of Power Sources*, 1990, **29**, 1–2, 3.

[9]    C. Stone and A.E. Morrison, *Solid State Ionics*, 2002, **152–153**, 1.

[10]   B. Verspagen, *Advances in Complex Systems*, 2007, **10**, 01, 93.

[11]   F. Barbir in *PEM Fuel Cells: Theory and Practice*, Academic Press, Cambridge, MA, USA, 2013, pp 4.

[12]   https://commons.wikimedia.org/wiki/File:Apollo_SM_fuel_cell.jpg.

[13]   https://commons.wikimedia.org/wiki/File:PureCell%C2%AE_System_Model_400.jpg.

[14]   https://commons.wikimedia.org/wiki/File:Hydrogenics_fuel_cell_Bus.JPG.

[15]   https://commons.wikimedia.org/wiki/File:Mercedes-Benz_F- Cell_WAS_2010_8926.JPG.

[16]   https://commons.wikimedia.org/wiki/File:Toyota_L%26F_7FB25_Fuel_Cell_Forklift_at_Eco-Products_2015.jpg.

[17]   https://commons.wikimedia.org/wiki/File:Das_von_Brennstoffzellen_angetrieben_Flug zeug_Antares_-_The_fuel- cell_powered_aircraft_Antares_(14050859619).jpg.

[18]   https://commons.wikimedia.org/wiki/File:Suzuki_Burgman_Fuel_Cell_Scooter.jpg.

[19]   M.Y. Kariduraganavar, R.K. Nagarale, A.A. Kittur and S.S. Kulkarni, *Desalination*, 2006, **197**, 225.

[20]  A. Kraytsberg and Y. Ein-Eli, *Energy & Fuels*, 2014, **28**, 12, 7303.

[21]  B. Smitha, S. Sridhar and A.A. Khan, *Journal of Membrane Science*, 2005, **259**, 1, 10.

[22]  J.O'M. Bockris and S. Srinivasan in *Fuel Cells: Their Electrochemistry*, McGraw-Hill, New York, NY, USA, 1969.

[23]  M. Rikukawa and K. Sanui, *Progress in Polymer Science*, 2000, **25**, 10, 1463.

[24]  S. Banerjee and D.E. Curtin, *Journal of Fluorine Chemistry*, 2004, **125**, 1211.

[25]  A.K. Pabby, S.S.H. Rizvi and A.M.S. Requena in *Handbook of Membrane Separations: Chemical, Pharmaceutical, Food, and Biotechnological Applications*, Eds., A.K. Pabby, S.S.H. Rizvi and A.M.S. Requena, CRC Press, Boca Raton, FL, USA, 2015, p.583.

[26]  A.B. LaConti, H. Liu, C. Mittelsteadt and R.C. McDonald, *ECS Transactions*, 2006, **1**, 199.

# 2 Proton-conduction membranes: Requirements, challenges and materials

## 2.1 Overview of membrane materials and concepts

The comprehensive commercial feasibility of systems such as fuel cells cannot be achieved unless costs are optimised for every component within such a system. The proton-conduction membrane (PCM), which is traditionally a polymer electrolyte membrane (PEM), forms the 'heart' of the fuel cell system. Primarily responsible for proton conduction (conduction of hydroxide ions in the case of alkaline fuel cells) and separating the reactions and reaction gases generated at the sides of the anode and cathode. It also serves various other essential functions, as mentioned briefly in Chapter 1. According to United States (US) Department of Energy (DOE) reports, the membrane cost for an 80-kW transportation fuel cell system operating on direct hydrogen in 2015 was $20/m^2$, whereas the target for 2020 is $10/m^2$ [1]. Furthermore, according to the 2016 report on the costs of fuel cell systems by the US DOE, PEM contribute to 26, 14 and 8% of the costs of fuel cell stacks at a rate of 1,000, 100,000 and 500,000 units of 80 kW produced per year, respectively [2]. The units produced in the early years must be restricted until the market is developed and ready to 'absorb' the technology. Therefore, despite this successive reduction in the PEM contribution to the costs of fuel cells, the urgent need to achieve further cost reductions while improving the PEM/ PCM performance persists, and forms one of the driving factors towards achieving an economically competitive fuel cell system.

Apart from high costs, the current commercial standard (Nafion®-based) PEM suffer from major drawbacks such as: a limited operating temperature (OT) range [OT <100 °C] for reliable use; loss of water content and proton conductivity at higher temperatures; dependence on water content (required for humidification) for proton transport; accelerated degradation at higher temperatures; degradation and chain scission in the presence of hydrogen peroxide ($H_2O_2$) and metal ions. PEM deterioration is one of the key factors responsible for loss of performance and degradation of fuel cells [1]. All these reasons necessitate further research and development in PCM. Several approaches intended to improve PEM have been explored and a huge variety of polymers (chemically-modified side chains) as well as inorganic materials have also been investigated to produce novel PCM. Figure 2.1 details the types of polymers and other inorganic materials explored for proton-exchange membrane fuel cell (PEMFC) and direct methanol fuel cell (DMFC) applications. However, before discussing the various membrane materials and preparation methods, a fundamental understanding of the functions which a PCM needs to deliver and the challenges it faces in this process due to the complex environment within an operating fuel cell is needed. These have been discussed in the next section.

https://doi.org/10.1515/9783110647327-002

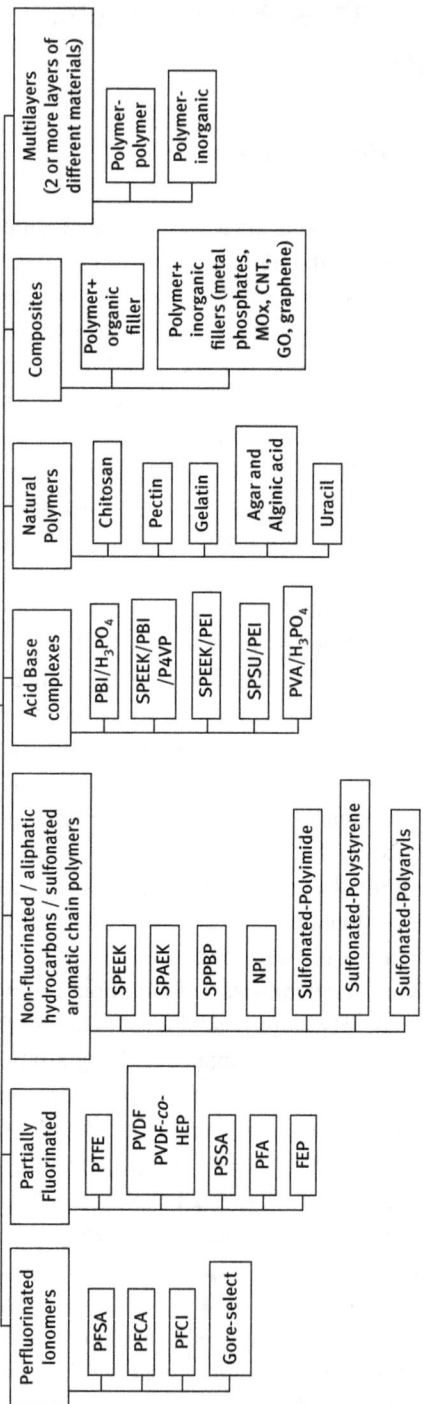

**Figure 2.1:** Different materials used for PCM in fuel cells (CNT: carbon nanotubes; CS: chitosan; FEP: poly(fluoroethylene-*co*-hexafluoropropylene); GO: graphene oxide; HEP: hexafluoropropylene; NPI: naphthalenic polyimide; P4VP: poly(4-vinylpyrrolidone); PBI: polybenzimidazole; PEI: polyetherimide; PFA: poly(tetrafluoroethylene-*co*-perfluorovinyl ether); PFCA: perfluorocycloalkene; PFCI: perfluorocarboxylated ionomer; PFSA: perfluorosulfonic acid; PSSA: polystyrene sulfonic acid; PTFE: polytetrafluoroethylene; PVA: polyvinyl alcohol; PVDF: polyvinylidene fluoride; SPAEK: sulfonated poly aryl ether ketone; SPEEK: sulfonated polyether ether ketone; SPPBP: Sulfonated poly(phenoxy benzoyl phenylene); phenylene); and SPSU: sulfonated polysulfone).

## 2.2 Requirements and challenges

Proton conduction in membranes in PEMFC [including DMFC and intermediate temperature proton-exchange membrane fuel cells (IT-PEMFC)] is dependant primarily on water molecules acting as the proton carriers. The hydration behaviour of PEM is, therefore, of paramount importance in determining their proton-conduction behaviour at different OT and humidity conditions.

### 2.2.1 Water uptake and proton conduction

Water content in membranes is defined in terms of grams of water per gram of dry polymer weight or in terms of the number of water molecules per sulfonic acid group (i.e., water content = $nH_2O/nSO_3H$) in the polymer (in case of PFSA membranes). A fully-hydrated Nafion® membrane is considered to hold ≤22 water molecules per sulfonic acid group. The water uptake of a membrane is dependent on the membrane material (number of sulfonic acid groups), membrane thickness, OT as well as the pre-treatment of the membrane. Membranes dependent on water for proton conduction require pre-treatments (also referred to as 'activation'), such as boiling in water and dilute acid, for eliminating contaminants and activation before they can be utilised to full strength. A dry or untreated membrane shows significantly lower water uptake. Also, if the membrane dries out at increased temperatures (for example, close to 100 °C) during operation, the rehydrated membrane may not achieve the water-content values similar to that of the initially hydrated membrane. This will depend on the membrane material [glass transition temperature ($T_g$)] and the morphological changes the membrane may go through at certain increased temperatures. For example, Dow membrane has been demonstrated to achieve its original water-uptake capacity (after dehydrating at 105 °C) upon rehydration at 80 °C, unlike Nafion® [3].

In low-temperature fuel cells, water for membrane humidification is supplied in the form of humidified hydrogen and oxygen gases. Therefore, within the operating PEMFC/DMFC system, the water-uptake behaviour of the membrane is also dependent on its ability to adsorb water from the vapour phase. In general, water uptake in PEM in the vapour phase is lower than water uptake in the liquid phase (vapour-phase water uptake for Nafion® leads to water content = 14). This phenomenon is called 'Schroeder's paradox', and occurs because, in the vapour phase, water adsorption takes longer due to the fact that it first requires condensation of the water molecules within the polymer (possibly at the hydrophobic backbone) before it can be adsorbed. The theory is supported (at least in part) by the water-uptake behaviour in the vapour phase, which has been identified to be in two distinct stages: low vapour activity and high vapour activity. During low vapour activity, water uptake takes place *via* solvation by ions in the membrane; in high-vapour activity, water fills within the pores and causes the membrane to swell [4]. Membrane swelling resulting in dimensional changes (usually ≤10%) must be considered during designing and installation of the cell and membrane inside the fuel cell.

As mentioned above, water content is the primary driver of proton conduction. However, the membrane must first fulfil two basic structural requirements for it to be eligible for ionic conduction (protons being the positively charged ions) well before water uptake can play its part:
1. Existence of fixed charged sites, and
2. Availability of free volume.

The presence of charged sites with charges opposite to that of the moving ions (i.e., fixed negative charge sites in case of moving protons, and *vice versa* in the case of alkaline fuel cells) serve as momentary 'reception and release centres' for the moving ions while ensuring a net charge balance across the polymer. The free volume (which refers to the pores present due to the intrinsic spatial organisation of the porous polymer structure) facilitates the movement of ions (protons) throughout the polymer. Small-scale structural vibrations also increase with increase in free volume, which can assist the physical transfer of ions from one charged site to another. Free volume, which is a characteristic of polymer structures, is responsible for the high ionic/proton conductivity exhibited by them as opposed to solid-state materials such as ceramics [5]. Technically, the ionic conductivity of the PCM used in fuel cell application must be 50–200 mS/cm for a reasonable operation. Also, the area specific resistance must be ≤0.15 Ohm.cm$^2$ [6].

This information brings us to the proton conduction in hydrated membranes. Proton conduction in hydrated membranes is explained with the help of two mechanisms: vehicular (or diffusion) and Grotthuss (or proton hopping). The vehicular mechanism is reliant on the free volume (readily available in PEM). In this case the hydrated proton, or the hydronium ion, diffuses through the aqueous medium along the free volume due to the electrochemical difference. These hydronium ions also tend to 'drag' water molecules with them ('electro-osmotic drag'). In the Grotthuss mechanism, the proton essentially hops from one protonated ionic site to another. For example, a proton generated at the anode attaches itself to a water molecule to form a hydronium ion, while another proton from the newly formed hydronium ion detaches itself and hops onto another nearby water molecule. The schematic in Figure 2.2 shows the two mechanisms. In PFSA membranes, it is generally accepted that the vehicular mechanism occurs in the case of low hydration and limited water content whereas the Grotthuss mechanism dominates in fully hydrated conditions.

(a)

Sulfur   Oxygen   Hydrogen

(b)

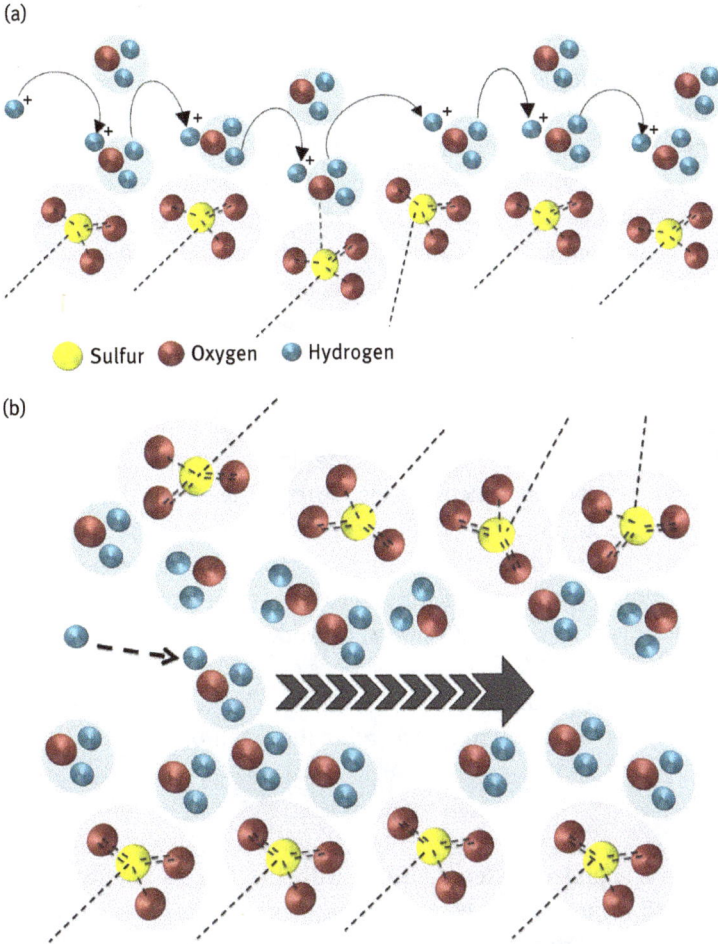

**Figure 2.2:** The two water-based proton-conduction mechanisms in membranes a) Grotthuss and b) vehicular (schematic).

### 2.2.2 Relationship between proton conduction, water transport and water concentration gradient in polymer membranes

As mentioned above, proton conduction *via* the vehicular mechanism results in electro-osmotic drag (i.e., water molecules are dragged along with the moving proton from the anode to the cathode side).

The flux of water electro-osmotic drag, 'N' is given as:

$$N_{H_2O,\,drag} = \varepsilon(\lambda)\frac{j}{F} \tag{2.1}$$

Where, '$\varepsilon$' is the electro-osmotic drag coefficient, 'j' is the current density in A/cm$^2$, and 'F' is the Faraday constant.

The amount of water being dragged per proton cannot be independent of the amount of water present in the membrane. Hence, '$\varepsilon$' is a function of the water content '$\lambda$'. The electro-osmotic drag coefficient, defined as the number of water molecules being dragged per proton, has been studied and different values have been reported for it. La Conti and co-workers used the method of passing current through the membrane and monitoring the level of water in a column and reported $\varepsilon = 2$–$3$ for $15 \leq \lambda \leq 25$, whereas Zawodzinski and co- workers reported values of 2.5 for a fully hydrated and immersed Nafion$^®$ 117 [3, 7, 8]. Later, Zawodzinski and co-workers, using the method of Fuller and Newman (the electrochemical method of measuring the potential difference arising across a membrane due to water activity gradient) reported the value of the coefficient to be 1.0 for $1.4 \leq \lambda \leq 14$ for a membrane (Nafion$^®$ 117) equilibrated in water vapour and 2.5 for membranes immersed in liquid water [9, 10]. Electro-osmotic drag is also dependent on the current density. As the latter increases, increasing numbers of protons move from the anode to cathode to further increase the flux of electro-osmotic drag, thus making the anode side of the membrane more prone to drying out. *In situ* investigation of water transport in PEMFC through water- balance experiments carried out by Yan and co-workers demonstrated that the electro-osmotic drag ranged from 1.5 to 2.6 under various operating conditions [11]

Apart from the water being dragged to the cathode side, water is also being electrochemically generated at the cathode, according the Equation 2.2 below:

$$N_{H_2O,\,gen} = \frac{j}{2F} \tag{2.2}$$

The electrochemically generated water, together with the electro-osmotic drag, eventually gives rise to a large water concentration gradient. This results in the back diffusion of water. The diffusivity of water, as studied for PFSA/polymer membranes, is a function of '$\lambda$'. Hence, the rate of back diffusion of water is given as:

$$N_{H_2O,\,diff} = D(\lambda)\frac{\Delta C}{\Delta Z} \tag{2.3}$$

Where 'D' is the coefficient of water diffusion in the polymer and $\Delta c/\Delta z$ is the water concentration gradient along the z-direction of the membrane [i.e., through (perpendicular to the plane of) the membrane] [5].

These three forces working together are responsible for the movement and transport of water in the membrane. The next flux of water in the membrane is, thus, a complex function of '$\lambda$'. Moreover, water may also be hydraulically 'pushed' in a particular direction in case of any pressure gradients between the anode and cathode. There is also a 'thermal-osmotic' drag which has not been very well investigated. This is because it does not seem to play a major part in back diffusion but it assists in maintaining the water balance to some extent [12]. The back diffusion of water can help with another issue: the anode side of the membrane drying out due to electro-osmotic drag. Back diffusion of water can counter this problem to keep the membrane hydrated. However, whether this phenomenon can be useful for maintaining hydration in thicker membranes is controversial. Earlier studies suggested that, in the case of thicker membranes, back diffusion could not ensure hydration of the membrane through its entire thickness whereas thinner membranes could be kept hydrated through this process [13, 14]. However, recent *in situ* studies carried out using neutron imaging suggest that, even with thicker membranes (e.g., Nafion® 117), back diffusion of water is evident [15]. Studies using *in situ* experiments are discussed elaborately in Chapter 5.

### 2.2.3 Challenges – ionic transport *versus* membrane thickness

In comparison with electronic conductivity in metals, ionic (protonic or cationic) conductivity in membranes is intrinsically lower by orders of magnitude. In order to decrease the ionic resistance, the path traversed by ions when moving across the membrane from the anode to cathode must be the shortest possible. The only way to achieve this is to reduce the membrane thickness. Thinner membranes are, therefore, preferred. As seen above, ionic transport suffers due to relatively thicker membranes. Also, thinner membranes may be less likely to suffer from the anode drying that occurs due to electro-osmotic drag.

However, thinner membranes are prone to various other issues:
- Reduced mechanical strength – making them susceptible to pin holes and, later, membrane failure.
- Electrical shorting – leading to sudden breakdown of the fuel cell.
- Dielectric breakdown – dielectric breakdown occurs if the membrane is so thin that the electrical field across it surpasses the dielectric breakdown field of the membrane material.
- Fuel crossover – as the membrane thickness decreases, crossover of reactant gases is easier, leading to various losses and other issues.

On the other hand, thicker membranes not only have higher ionic resistance but also contribute to the contact resistance [with the catalyst and gas diffusion layer (GDL)], net resistance and related losses in the fuel cell. Thus, the optimum thickness and the choice of membrane material are extremely important [5].

### 2.2.4 Challenges specific to application of fuel cells

The intended application of the fuel cell also gives rise to specific demands in terms of the characteristics and performance of the membrane. For example, membranes used in fuel cells for stationary applications must be more durable as compared with automotive applications. While ideally, membranes that can deliver good performance at higher OT and low relative humidity (RH) conditions are best for use in fuel cells for automotive applications. The current Nafion® and PFSA membranes largely determine the OT and humidity requirements in an operating fuel cell system. PFSA membranes have a $T_g$ <120 °C, so they are predisposed to hole formation and creep at temperatures >80 °C. Therefore, there is an increasing interest in exploring membranes that can operate close to 100–120 °C and low RH (~50%). Existing fuel cell systems also suffer from the drawback of additional cooling requirements as compared with internal combustion engine (ICE) automotives. Fuel cells reject only 10% of the heat with the exhaust gases as opposed to ICE, which convert >60% of their fuel energy into heat, making it easier to remove waste heat [6]. One reason for this is that the PEMFC stack temperature (usually 60–80 °C) is well below that of ICE (~120 °C). All these reasons result in the requirement of elaborate cooling and humidification systems, and add to the costs of fuel cell automotive and stationary power systems. The search for PCM which can easily operate ≤120 or ≤200 °C and at zero or low RH would, therefore, significantly improve the logistics of automotive and stationary PEMFC, and make them more competitive against ICEs.

### 2.2.5 Specific challenges for membranes in proton- exchange membrane fuel cells

The variable temperature and humidity conditions along with variable load changes give rise to a highly dynamic environment inside an operating fuel cell. Factors such as continual start-up and shutdown cycles, fuel shortage, impure gas feed, operation under low-RH conditions, gas crossover due to intrinsic porosity of PEM, and electrocatalyst loss contribute actively to membrane degradation. The inevitable pressure and concentration gradients give rise to the convection and diffusion of gases from one side to another. This leads to fuel dissipation because of a direct reaction between the hydrogen and oxygen inside the membrane. Such direct reactions inside the membrane also generate local 'hotspots' as well as radical chemical species. These radicals along with stray, lost electrocatalyst particles, react further with the membrane, leading to membrane thinning, pinhole formation, degradation and, eventually, membrane failure. The mechanical strength of the PEM is also put to test continuously with variable hydration levels that lead to swelling and contraction of the membrane.

Hydrogen crossover, owing to the size of the molecule, is a commonly faced issue. The small size of the molecule makes it nearly impossible to prevent it from crossing over.

### 2.2.6 Specific challenges for membranes in direct methanol fuel cells

DMFC, as a branch of PEMFC, gained popularity because of the portability of methanol (MeOH) as the fuel. The relative ease of handling, storage and transportability of MeOH as compared with hydrogen, along with its high energy density (6,100 Wh/kg at 25 °C) and the fact that it remains a liquid over a wide temperature range, make it a much more convenient alternative fuel in comparison with hydrogen, where challenges such as compression/liquefaction for storage and transportation require significant infrastructural inputs. Production of bio-methanol is gaining importance because it can be produced from industry, household and animal wastes. These properties render DMFC a highly favourable powering system for small and mobile electrical devices (laptops, smartphones, digital cameras, other handheld electronic devices). However, the membranes used in DMFC have to face some additional challenges. The biggest drawback faced by DMFC and their membranes is the crossover of MeOH. The hydration-dependant PFSA membranes achieve highest proton conductivity at 100% RH. MeOH is soluble in water and, therefore, permeates through the PEM, into the cathode, resulting in a competitive (mixed potential) reaction at the cathode. Decreasing the water uptake of the membrane is not an alternative because it will be deleterious to the proton conductivity of the membrane. As such, relatively thicker membranes (e.g. Nafion® 117) are used in state-of-the-art DMFC systems [16].

Other approaches to reduce and minimise MeOH crossover have been investigated. These include the use of inorganic filler particles, which are discussed later in this chapter.

## 2.3 Types of polymer electrolyte membranes

### 2.3.1 Perfluorosulfonic acid membranes

PFSA membranes are the most widely used and studied for PEMFC and DMFC systems. First developed during the 1960s by Dr. Walther Grot from DuPont, for application in chlor-alkali production, PFSA have been improved significantly and are produced by many other companies now. Nafion®, the first and most popular PFSA, is a modified version of another DuPont ionomer, Teflon™. Nafion® consists of a Teflon™/polytetrafluoroethylene (PTFE) backbone with an added side chain and a pendant sulfonic acid group. The hydrophobic PTFE backbone imparts chemical and thermal stability to the ionomer whereas the hydrophilic sulfonic acid group is in charge of hydration (i.e., water uptake and ion-exchange capacity of the membrane). This sturdy backbone provides the PFSA membrane with good mechanical stability and good temperature range for operating in various applications. The sulfonic groups can organise themselves into hydrophilic channels if exposed to water molecules. As more water becomes available around the sulfonic groups, the water clusters grow in size and cause the Nafion® polymer to separate and form into hydrophobic and hydrated hydrophilic regions that form channel-like structures, enabling water transportation and limiting swelling of the membrane upon water adsorption. However, there is a percolation threshold below which Nafion® only acts as an insulator as the small water clusters remain isolated. As water content increases, channels are formed and become broader. At this stage, diffusion-based proton conductivity begins as the activation energy is quite high for hop-on proton transfer. As a saturated stage is reached inside Nafion®, Grotthuss-type conductivity is achieved which, until now, was impeded by the relatively strong electrostatic forces acting between the hydronium ions and sulfate groups [6, 17–19]. PFSA demonstrate proton conductivity of 0.1–0.2 S/cm if operated at 80 °C under hydrated conditions. The various other commercial PFSA membranes vary in the length of their side chain (Chapter 1) and preparation methods. Gore®, Aciplex™ and Flemion™ membranes are mechanically reinforced membranes. The reinforced membranes show better tensile strength, less shrinkage during drying, and less creep under stress. Different-thickness and reinforced varieties of Nafion® have also been introduced [6, 17].

However, irrespective of their dominance in the commercial membranes sector (especially fuel cells), these membranes suffer from several drawbacks. Most of these drawbacks stem from the water dependence of PFSA membranes for proton conductivity. Reliance on water-filled channels for proton conductivity means PFSA struggle for operation <0 or >100 °C. Use of pressure in the PEMFC system allows water to stay in liquid form at slightly higher temperatures so that the OT range of the system can be changed, but the other physical limitation of PFSA approaching its $T_g$ (120 °C) cannot be ignored: as the OT increases the chemical and mechanical strength of PFSA decreases. At lower temperatures (≤80 °C), the membranes are chemically sturdy even in the presence of 30% $H_2O_2$, but the combined presence of 'alien'/stray metal ions and $H_2O_2$ is known to cause havoc by catalysing the decomposition of the polymer chains, especially the ends of the carboxylic acid groups [20]. The high solubility of MeOH in water means it finds easy access, permeating through the water channels formed inside the hydrated PFSA. The reinforced membrane structures increase mechanical stability and MeOH tolerance but compromise on proton conductivity and water vapour permeability [21]. The economics of the ionomers is also very important. The monomers for manufacture of PFSA membranes are expensive to synthesise due to multi-step processes and the expertise required for safe handling of dangerous chemicals. Moreover, the limited usage of the perfluorosulfonyl fluoride ethyl propyl vinyl ether monomer means limited volume production aimed especially for the fuel cell industry [6, 17].

All these drawbacks and limitations have encouraged a wide variety of non-PFSA membranes to be investigated over the last few decades: the following sections discuss these membranes. A selected summary of various types of alternative membranes reported in recent times is shown in Figure 2.1.

### 2.3.2 Partially fluorinated

Partially fluorinated membranes consist of composites of fluoride-based polymers [e.g., PTFE, polyvinylidene fluoride (PVDF)] with various aromatic polymers. These are prepared using methods such as reinforcement, grafting and blending. These polymers offer potentially excellent mechanical and chemical stability. FuMa-Tech and Ballard produce partially fluorinated membranes. Some of the membranes produced by these companies include the FKH series by FuMa-Tech as well as BAM1G [series of sulfonated poly(2,6-diphenyl-1,4-phenylene oxides)], another series based on sulfonated polyaryl ether sulfone and BAM3G (sulfonated membranes based on α,β,β-trifluoro-styrene monomers, and a series of substituted trifluoro-comonomers) by Ballard [22, 23]. Other partially fluorinated polymers reported and studied include partially fluorinated disulfonated polyarylene ether benzonitrile copolymers. Many of these have displayed reduced MeOH permeability and better durability than Nafion® in *ex situ* studies [24, 25].

Radiation grafting process to synthesise partially fluorinated membranes entails an electron beam or irradiation step which produces radical sites in the membrane (PFSA, partially fluorinated and non-fluorinated). Then, selected segments can be grafted onto the base polymer sites. Following this, the membranes are swollen with suitable polymer solutions of network-forming compounds and an interpenetrating polymer network is achieved by further heating. The process is completed by sulfonation. This process offers low cost, simple reactions, and the potential to form crosslinked material in its final form.

### 2.3.3 Sulfonated hydrocarbon

These membranes include non-fluorinated aliphatic and aromatic hydrocarbons, which are sulfonated to add the hydrophilic ends that enable them to conduct protons. The use of hydrocarbons in polymer membranes as polymer backbones offers several advantages over current state-of-the-art membranes: cost benefits, ease of recycling, and the convenience of attaching additional pendant groups on polar sites. The polar groups also enable hydration at high temperatures because they demonstrate high water uptake over a broad range of temperatures [18]. The hydrocarbons, however, show low ionic conductivity compared with perfluorinated polymers <100 °C. Interestingly, they show performances comparable with that of Nafion® >150 °C [26, 27].

An important parameter, apart from the method of sulfonation/sulfonating agent and choice of polymer combination, is the degree of sulfonation. As the degree of sulfonation increases, chemical stability and mechanical strength are compromised. Moreover, excessive sulfonation results in enhanced hydrophilicity that can eventually lead to dissolution of the sulfonated polymer in water. Depending on the intrinsic properties of the hydrocarbon backbone, a variable degree of sulfonation can be achieved safely to attain an optimum combination of proton conductivity as well as chemical and mechanical stability.

### 2.3.4 Acid–base complexes

Acid–base complexes are economically viable, high-proton conductivity systems. These systems work independently of the hydration level because proton conduction here does not require the presence of liquid water. As the name suggests, these complexes consist of acid and base components. Essentially, an acid component is introduced into a basic/alkaline polymer base to impart proton conduction. Polymerisation takes place between the diamine of the polymer and acid. The various ionic crosslinking and hydrogen-bonding interactions that take place between acid and base polymers contribute markedly in controlling membrane swelling without loss of flexibility. The most commonly studied acid–base complex is phosphoric acid ($H_3PO_4$)-doped polybenzimidazole (PBI). The latter is an aromatic hydrocarbon in which proton conductivity is introduced by doping with $H_3PO_4$ [19, 28]. PBI is also mechanically and thermally stable ($T_g$ = 430 °C), making these membranes ideal for high-temperature low-RH operations. Proton conduction in PBI is believed to take place *via* the Grotthuss mechanism whereby the free acid vehicle involves $H_3PO_4$. PBI displays performance similar to that of Nafion$^®$ ($\approx$0.2 S/cm) even with 40% RH at 80 °C [29]. The high-temperature operation allows fuel cells to be more tolerant to impurity while the heat generated is also easier to dissipate, which can be utilised effectively in combined heat-and-power devices.

Other acid–base complexes reported include polyetherimide (PEI), and sulfonated polymers such as sulfonated polyphthalazinone ether nitrile ketone, sulfonated polyphthalazinone ether sulfone ketone and poly(4-vinylimidazole). Sulfonated polyether ether ketone (SPEEK)/ PBI membrane composites have also been studied.

The key consideration in synthesising an acid–base system is the level of doping, which is measured in terms of $H_3PO_4$ mole percent per repeating unit of the polymer. As the doping level increases, the gap between the acid sites decrease, thereby increasing conductivity. However, mechanical properties deteriorate as the doping level increases. Another shortcoming linked with acid–base membranes is the possibility of acid leaching at high acid-doping levels due to thermal as well as mechanical instability, a subject that needs further investigation. Blends with specially designed bases with nano- and micropores are being explored which demonstrate promising prospectives towards minimising or eliminating these problems in the future [30–34]. One such example is the work reported by Hazarika and Jana, in which they prepared blends of PBI and poly(1-vinyl-1,2,4-triazole) (PVT) using solution processing which were stable ≤300 °C and also allowed higher loading of $H_3PO_4$ (Figure 2.3) [30].

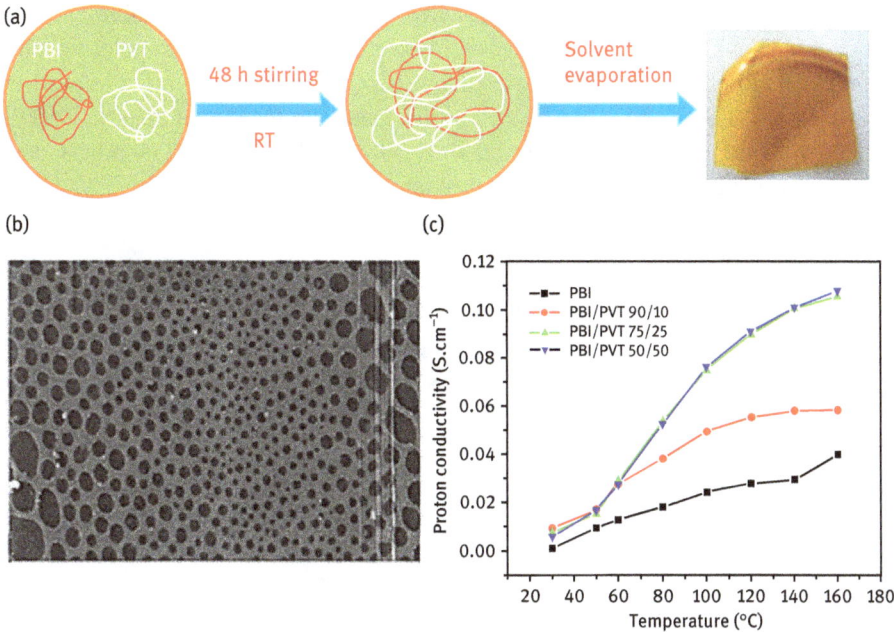

**Figure 2.3:** a) Preparation of polymer blends; b) scanning electron microscopy image of a 90:10 PBI/PVT blend at a scale of 1 µm collected at 300 kV; and c) proton conductivity of the different PBI/PVT blends compared with PBI (RT: room temperature). Reproduced with permission from M. Hazarika and T. Jana, *ACS Applied Materials & Interfaces*, 2012, **4**, 5256. ©2012, AmericanChemical Society [30].

### 2.3.5 Natural polymers

Natural polymers can be low cost, lightweight, safer to synthesise and handle, and biodegradable alternatives to synthetic polymers. Several polysaccharides and protein-based polymer structures are considered to be promising candidates due to their abundance in nature and variety of chemical compositions and structures they can offer. Natural polymers are already used widely in the food industry so there is plenty of knowledge, expertise and skill to translate for further experimentation in fuel cell membranes. Natural polymeric materials such as CS, agar-agar, pectin, gelatin, alginate and hydroxyethyl cellulose have been popular choices in recent years for studies on PCM for application in fuel cells as well as other electrochemical devices [35–38].

Most natural polymers offer proton conductivities of $10^{-4}$ to $10^{-5}$ S/cm, but combining them with other polymers and materials can help improve proton conduction. CS is derived from chitin (a polysaccharide found in crustaceans and insect shells) and is the most explored among natural polymers for PEM applications. CS, apart from the above-listed benefits of a natural polymer, offers other advantages. It is hydrophilic and insoluble in water as well as most organic solvents. It is inert and has good chemical and thermal stability. The presence of free amino and hydroxyl functional groups in the CS backbone offer immense scope for 'tailoring' and chemical modifications [36, 37]. These unique properties have generated interest in the study of CS-based hybrid and composite materials for membranes in DMFC, PEMFC, alkaline fuel cells and microbial fuel cells [39, 40]. Table 2.1 details studies that have used natural polymer composites for DMFC membranes.

**Table 2.1:** Natural polymer-based membranes for DMFC.

| Membrane | Proton conductivity (S/cm) | Methanol permeability (cm$^2$/s × 10$^{-6}$) | Active component for proton conduction | Ref |
| --- | --- | --- | --- | --- |
| CS–silica–PVA hybrid polyelectrolyte | 0.0531 | 1.26 | Silica | [41] |
| CS–PVP | 0.019 | 0.092 | PVP | [42] |
| PVA–sodium alginate/ hetropolyacids mixed matrix membrane | 0.024 | 0.168 | Heteropolyacids | [43] |
| Caesium phosphotungstate salt–CS composite | 0.006 | 0.56 | Caesium phosphotungstate salt | [44] |
| PVA–amino acid–TiO$_2$ hybrid nanocomposite | – | – | Amino acid-functionalised titania | [45] |
| Sodium alginate/ sulfonated GO | – | – | Sulfonated GO | [46] |
| Sodium alginate/ [poly(ether-block-amide)/sulfonated gluteraldehyde] | 0.0067 | 0.0925 | Poly(ether-block-amide) | [47] |

PVP: Polyvinylpyrrolidone

## 2.3.6 Composites

Composite membranes are usually defined as polymer membranes modified with organic and inorganic filler materials. The filler materials are chosen for a very specific motive to meet definite ends, such as increasing membrane tortuosity to slow/decrease MeOH crossover (Figure 2.4), self-hydration, increasing mechanical strength, and increasing proton conductivity. A wide variety of fillers have been studied which can be classified broadly based on the type of materials: polymeric, metal oxides [titanium dioxide (TiO$_2$), silicon dioxide, zirconium dioxide (ZrO$_2$), sulfonated ZrO$_2$, zirconium titanium phosphate], carbon nanomaterials [CNT, graphene, graphite oxide (GO)] and their functionalised (sulfonated, phosphonated, carboxylated) counterparts. Polymeric fillers can be seen simplistically as polymer–polymer composites

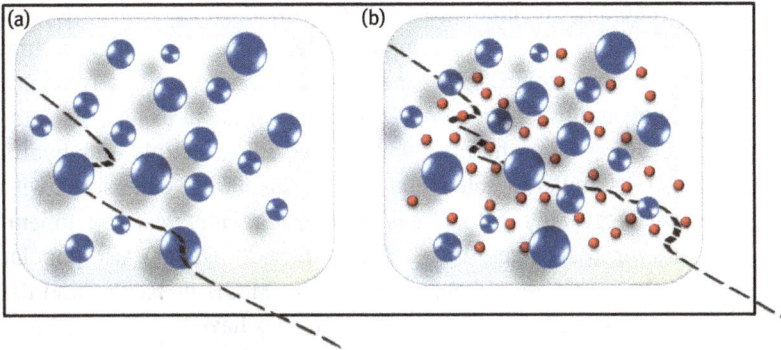

**Figure 2.4:** Schematic representative showing the difference in fuel path between membranes with and without fillers, representing the fuel path (dashed line) in a) the membrane with no filler and b) the composite membrane with filler particles (small circles). Reproduced with permission from C.M. Branco, S. Sharma, de Camargo Forte and R. Steinberger-Wilckens, *Journal of Power Sources*, 2016, **316**, 319. ©2016, Elsevier [48].

**Table 2.2:** Performance of zeolite-based membranes in alcohol-based fuel cells.

| Polymer | Zeolite | Method | Alcohol permeability (cm²/s) | Proton conductivity (mS/cm) | Selectivity (s/cm³) | Ref |
|---|---|---|---|---|---|---|
| Nafion® | N/A | N/A | $13.1 \times 10^{-7}$ | 95.6 | $7.3 \times 10^4$ | [69] |
| Nafion® | Fe-silicate-1 | Immersion in zeolite solution | $8.40 \times 10^{-9}$ | 3.8 | $4.5 \times 10^5$ | [70] |
| PVA | MOR | Blending casting | $1 \times 10^{-7}$ | 100 | $10^5$ | [71] |
| Nafion® | LTA | Blending casting | $1.03 \times 10^{-6}$ | 35 | $3.4 \times 10^7$ | [72] |
| CS | ZSM-5 | Blending casting | $1 \times 10^{-6}$ | 20 | $2 \times 10^7$ | [64] |
| CS | LTA | Blending casting | $1.5 \times 10^{-6}$ | 17 | $1.13 \times 10^6$ | [64] |
| Nafion® | Acid-functionalised zeolite beta | Blending casting | $3.63 \times 10^{-6}$ | 67 | $1.84 \times 10^7$ | [73] |
| CS | Beta | Blending casting | $2.07 \times 10^{-7}$ | 14.1 | $6.81 \times 10^4$ | [74] |
| PVA | MOR | Blending casting | $9.97 \times 10^{-9}$ | 50.3 | $5 \times 10^6$ | [75] |
| SPEEK | Formaldehyde in X-type faujasite zeolites | Immersion in zeolite solution | $3.15 \times 10^{-7}$ | 29.1 | $9.2 \times 10^4$ | [69] |

LTA: Linde Type A
MOR: Mordenite
Reproduced with permission from I.G.B.N. Makertihartha, M. Zunita, Z. Rizki and P.T. Dharmawijaya in *Proceedings AIP Conference 2017*, AIP Publishing LLC, 2017, **1818**, 1, 020030-10. ©2017, AIP Publishing LLC [62]

and some of the sulfonated hydrocarbon membranes can be considered to be polymeric composites [48]. Some examples of composite membranes combining natural and synthetic polymers along with different metal oxides can be seen in Table 2.2, which shows natural polymer-based membranes. In terms of metal oxides, silica (as particles and nanotubes) has been the most widely used filler to explore its hydrophilic properties. Due to its ability to hold water it has been studied extensively with various polymers aimed especially at fuel cells operating at intermediate temperatures (100–120 °C) and for low-RH conditions. Silicate-based fillers are also popular for DMFC membranes because they increase the path for MeOH molecules and decrease MeOH crossover. Nanostructured $TiO_2$, ZrOx and CeOx have stirred more interest in recent years. Ketpang and co-workers prepared Nafion® composite membranes using mesoporous metal oxide nanotubes of oxides of titanium, zirconium and cerium by calcination of electrospun polyacrylonitrile nanofibres implanted with metal precursors to achieve efficient migration and retention of water in Nafion® membranes to enable low-RH performance (Figure 2.5). These composite membranes, when compared with commercial Nafion® NR-212, facilitated up to 4-fold higher-maximum power density at 18% RH and 80 °C [49].

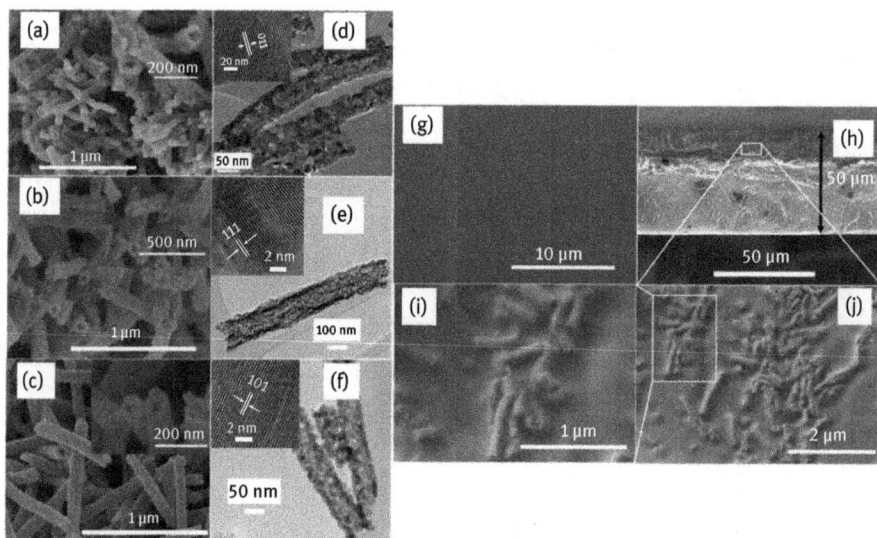

**Figure 2.5:** a–c) Field-emission scanning electron microscopy images of nanotubes of a) $TiO_2$, c), cerium(IV) oxide ($CeO_2$), (e) $ZrO_{1.95}$; d–f) transmission electron microscopy images and corresponding lattice fringes (inset) of b) $TiO_2$, (d) $CeO_2$, (f) and $ZrO_{1.95}$; g) top view of a Nafion®–$TiO_2$ membrane (thickness ≈50 μm), with 1.5 wt.% $TiO_2$; h) cross-section image of a Nafion®– $TiO_2$–1.5 membrane; and i and j) high-magnification cross-section images of a Nafion®–$TiO_2$–1.5 membrane. Reproduced with permission from K. Ketpang, K. Lee and S. Shanmugam, *ACS Applied Materials & Interfaces*, 2014, **6**, 16734.©2014, American Chemical Society [49].

Carbon-nanostructured materials, especially CNT, have also been used extensively as fillers for increasing mechanical strength and introducing tortuosity to limit MeOH crossover with encouraging results. Sulfonated carbon nanostructures have also been investigated and found to be useful in enhancing proton conduction while increasing mechanical strength for the membrane structure [50–52]. Graphene, GO and chemically-modified versions of GO (sulfonated-GO, amino-GO) have also garnered attention in recent years [53]. In a recent study, Shukla and co-workers reported the simultaneous unzipping and sulfonation of CNT to form sulfonated 'graphene ribbons' using a hydrothermal process, and used the nanoribbons as fillers in SPEEK-based membranes [54]. The as-prepared membranes with 0.1 wt% sulfonated graphene nanoribbons demonstrated high durability and negligible fuel (MeOH) crossover during a 175-h operation as reported by the authors. However, results have been conflicting for graphene (and GO), which is most likely due to inappropriate comparisons being made between materials of different purity, sheet size and number of layers (in the case of graphene); and different synthetic methods, oxidation levels, i.e., C:O ratio, type of oxygen groups (hydroxyl, carboxyl, carbonyl, epoxide) and their ratios, purity levels and sheet size, which can vary depending upon starting graphite and oxidation process (in the case of GO). The lack of standard conditions and variable test settings in *ex situ* and *in situ* studies of these two materials further contributes to the inconsistencies in this respect.

Another class of fillers is clays. Clays are metal [sodium, potassium or aluminium] silicates consisting of tetrahedral silica and octahedral alumina sheets. The metal cations in montmorillonite (MMT) are replaced with hydrogen ions ($H^+$) and these layered structures have been reported to show ionic conductivities of $10^{-4}$ S/cm. The primary motive of including clays in PCM is to enhance mechanical strength, reduce gas (or MeOH) permeability, and improve dimensional stability (including preventing excessive swelling at high RH and high-temperature conditions) [55, 56]. The concept of polymer–clay composites goes back to the early-1990s, when Kojima and co-workers studied rubber–clay composites to investigate their effect on gas permeability for Toyota [57]. Since then, several studies have explored the combination of polymers and clays for use in DMFC [58, 59]. Further functionalisation of MMT has also been explored, such as sulfonic acid functionalisation to improve their ionic conductivity before incorporation into the polymer [60] (Figure 2.6). More recently, it has also been studied as filler in natural polymers, where MeOH permeability of $3.03 \times 10^{-7}$ cm$^2$/s and proton conductivity of $4.66 \times 10^{-3}$ S/cm could be achieved [61].

Zeolites have also garnered considerable attention for composite membranes over the years. Zeolites are microporous crystalline materials with well-defined structures. Naturally occurring zeolites are a class of minerals consisting of hydrated aluminosilicates (usually of sodium, potassium, calcium or barium). Zeolites such as mordenite, chabazite, clinoptilotite, 3A, 4A, 5A, 13X and HZSM-5 have been studied especially for DMFC applications. These can be readily subjected to continuous rehydration and

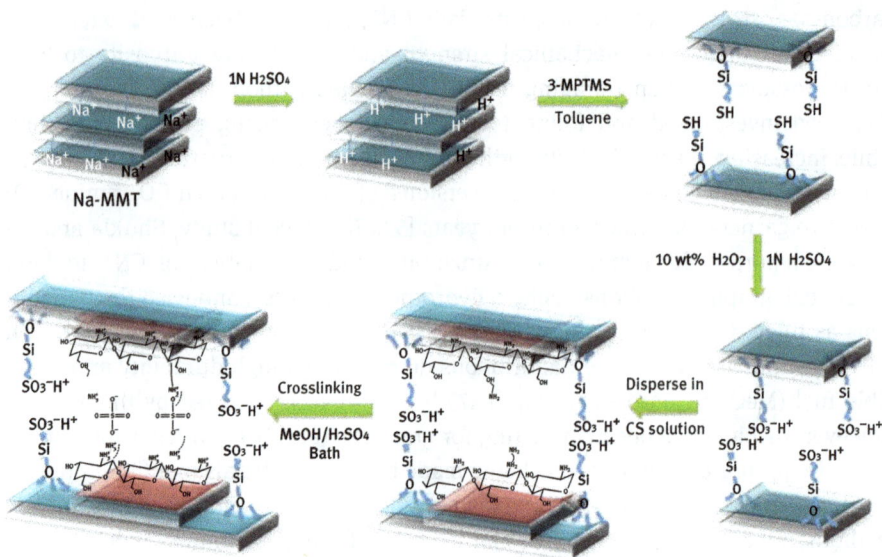

**Figure 2.6:** Schematic showing the sulfonyl functionalisation of sodium montmorillonite (Na–MMT) followed by preparation of a CS–MMT membrane using ionic crosslinking with sulfuric acid showing a CS chain with positively charged –(NH$_3$)$^+$ and negatively charged sulfuric acid –(SO$_4$)$^{2-}$ ionic interaction along with sulfonyl functionality of Na–MMT (3-MPTMS: 3-(mercaptopropyl)trimethoxysilane). Reproduced with permission from S.K. Nataraj, C. Wang, H. Huang, H. Du, Chen and K. Chen, *ACS Sustainable Chemistry & Engineering*, 2015, **3**, 302. ©2015, American Chemical Society [60].

dehydration cycles. Zeolites are also used in other fuel cell processes, including reforming of hydrogen gas, as well as the refining, purification and storage of fuel. As fillers for polymer-based membranes, they can reduce MeOH crossover significantly but also interfere with proton transport [62, 63]. Wang and co-workers reported an extensive study of zeolites as fillers in a CS matrix. They revealed that the MeOH permeability in a zeolite-filler composite membrane was not only dependent on the zeolite content but also on the zeolite particle and pore size and its hydrophilicity/hydrophobicity [64]. They demonstrated that hydrophilic zeolites could in fact increase MeOH permeability. Sasikala and co-workers showed that zeolite 13X modified with organo-sulfonic acid groups used as filler in sulfosuccinic acid–SPEEK membranes can serve the dual purpose of reducing MeOH crossover and providing additional ion-conducting groups. Zeolites have also been used on their own to replace traditional polymer membranes [65]. Other zeolites such as titanosilicates, faujasite and sepiolite have been utilised to enhance mechanical strength and water retention in Nafion® membranes [66–68]. Table 2.2 summarises some recent reported polymer–zeolite composites.

Apart from zeolites, another class of mesoporous materials prompting research interest is metal organic frameworks (MOF). MOF are coordination polymers consisting of metal centres linked to organic ligands. These offer advantages such as tunability of pore size and ease of functionalisation without compromising structural integrity along with extremely high surface area ($>3,000$ $m^2/g$). They have also exhibited endurance towards structural alterations upon temperature changes as well as pressure and pH variations [76]. These properties also make MOF-based composite polymer membranes well suited for gas storage and sensor applications. Hurd and co-workers developed a MOF, $Na_3$(2,4,6-trihydroxy-1,3,5-benzene trisulfonate), in which regular one-dimensional (1D) pores lined with sulfonate groups enabled proton conduction [77]. Proton conduction ($5 \times 10^{-4}$ S/cm at 150 °C) was modulated with the help of controlled loading of 1 H-1,2,4-triazole within the 1D pores. More studies are needed to develop deeper understanding of MOF filler-based composite membranes for proton conduction.

Ionic liquids (IL) have also gained significant interest due to their wide temperature and electrochemical potential range. IL are molten organic salts with melting points close to room temperature (RT). First identified in the early-1900s, they soon garnered considerable attention as a medium for nuclear-fuel reprocessing and as plasticising anion for polyethylene oxide (PEO)-based organic polymer electrolytes. With ions as their sole constituents and no solvents (unlike ionic solvents), these liquids consist of a large cation and charge-delocalised anions that result in weak interactions between ions and rendering it with a minimal inclination to crystallise. These features offer the advantages of extremely low volatility and vapour pressure, non-flammability, high thermal and electrochemical stability, and excellent ionic conductivities even in anhydrous conditions. Various IL have been developed based on the type of cation, such as dialkyl-substituted imidazolium, tetraalkylammonium, pyrrolidinium, piperidinium, and pyridinium cations. The ionic conductivities of these IL are known to be in the range of $1.0 \times 10^{-4}$ to $1.8 \times 10^{-2}$ S/cm. On the basis of their compositions, ILs have been classified into three categories: aprotic, protic and zwitterionic. Protic IL have a mobile proton located on the cation, making them suitable for proton-conducting electrolytes in fuel cells [78, 79]. The use of IL electrolytes would enable higher temperature ($>100$ °C) operations for alkaline and acid-based PEMFC and DMFC systems under anhydrous conditions because ion transport is dependent on the imidazole or amine group instead of the presence of water molecules. One current shortcoming of IL is the inability to form solid electrolytes. Several approaches, including immobilisation, polymerisation of the components, or gelification by a neutral macromolecule using different polymers (including PVDF, PVA, SPEEK, PBI) have been explored to develop IL-based composite PCM for PEMFC (and DMFC) [80–88]. These however, can lead to compromises on the mechanical strength of the membrane and other IL properties. Xu and co-workers reported the use of 3-aminopropyltriethoxysilane IL for an IL–GO/PBI composite membrane which, according to the authors, exhibited an appropriate level of proton conductivity when imbibed with $H_3PO_4$ at low-$H_3PO_4$ loading, thus promoting its use in fuel cells by avoiding acid leakage and material corrosion [88]. The IL–GO/PBI/$H_3PO_4$ composite membranes exhibited a proton conductivity of 0.035 S/cm per repeating unit of 3.5 at 175 °C. Table 2.3 compiles selected reports on IL-based membranes in recent years.

**Table 2.3:** Properties and performance of various IL-based membranes.

| Type of IL membrane | Proton conductivity (S/cm) | Methanol crossover (cm²/s) | Comments | Ref |
|---|---|---|---|---|
| Nafion® 112/ tetramethylammonium chloride | – | $4.21 \times 10^{-8}$ | 15 M Methanol used at RT | [89] |
| Nafion® 112/ phenyltrimethylammonium chloride | – | $3.89 \times 10^{-8}$ | 15 M Methanol used at RT | [89] |
| Nafion® 112/N -dodecyltrimethylammonium chloride | – | $1.56 \times 10^{-8}$ | 15 M Methanol used at RT | [89] |
| Nafion® 112/ hexadecyltrimethyl- ammonium bromide | – | $1.21 \times 10^{-8}$ | 15 M Methanol used at RT | [89] |
| Nafion® 112/1-N-butyl-3- methylimidazolium bis (trifluoromethanesulfonimide) | – | $5.16 \times 10^{-8}$ | 15 M Methanol used at RT | [89] |
| Nafion® 112/1-N-octyl-3- methylimidazolium bis (trifluoromethanesulfonimide) | – | $2.59 \times 10^{-8}$ | 15 M Methanol used at RT | [89] |
| Nafion® 112/methyl- tricaprylylammonium dicyanamide | – | $4.05 \times 10^{-9}$ | 15 M Methanol used at RT | [89] |
| [1-Ethyl-3- methylimidazolium] [bis (trifluoromethanesulfonyl) imide]/tetramethoxysilane/ methyl-trimethoxysilane/ trimethylphosphate | $5.4 \times 10^{-3}$ | H₂ crossover $=1.39 \times 10^{-12}$ mol/ cm s Pa | At 150 °C under anhydrous conditions, operational from −20 to 150 °C | [90] |
| SPEEK/1-butyl-3- methylimidazolium tetrafluoroborate | $1.04 \times 10^{-2}$ | – | At 170 °C under anhydrous conditions Interconnected ionic clusters observed on the composite membrane cast from dimethyl- formamide solutions | [91] |

**Table 2.3** (continued)

| Type of IL membrane | Proton conductivity (S/cm) | Methanol crossover (cm$^2$/s) | Comments | Ref |
|---|---|---|---|---|
| Hydroxyethylmethacrylate/ N-ethyl-N- methylpyrrolidinium (FH)$_{1.7}$F (1:9) | $8.19 \times 10^{-2}$ | – | Tested under non-humidified conditions. Obstruction of gas flow in gas diffusion electrode due to softening of the composite membrane at elevated temperatures | [92] |
| Sulfonated poly(styrene-isobutylene-styrene)/IL | 0.15 | $0.8 \times 10^{-6}$ | IL ($\approx$10 mol%) required for efficient proton conductivity | [93] |

Although this section has covered a broad landscape in terms of the materials used in composite membranes, it is not exhaustive. For further detailed information on the wide variety of composite membrane systems, the author recommends a recent review by Bakangura and co-workers [76].

# 2.4 Non-polymeric membranes

Materials other than polymers have also been studied over the last two decades, including solid acids and carbon nanomaterial (especially CNT and graphene)-based free-standing films as membranes.

## 2.4.1 Solid acid membranes

Solid acid membranes are a promising new class of membrane materials that offer a lot of potential for IT-PEMFC systems. Solid acids are intermediates between normal acids and salts whereby some of the hydrogen atoms in the traditional acid structure are replaced with cations to make them solid acids and behave as proton donors. These were first studied in the early-1980s and display high proton conductivity in anhydrous as well as humid environments, and high-temperature stability. Simply put, these are metal sulfates/phosphates. Caesium-based [caesium hydrogen sulfate ($CsHSO_4$) and caesium dihydrogen phosphate ($CsH_2PO_4$)] solid acids are suitable for 100–300 °C operation are the most widely studied as solid acids. They are also referred to as 'superprotonic electrolytes' because symmetric ($CsHSO_4$ and $CsH_2PO_4$) and asymmetric [$CsH(PO_3H)$] anions can form superprotonic phases. At low temperatures, other metal oxides (such as silica and titania) can be added to $CsHSO_4$ to improve the performance by 2–3 orders of magnitude [16]. However, sulfate-based solid acids react with hydrogen gas in the reducing anode environment to release hydrogen sulfide. As such, phosphate-based compounds are considered for long-term stability. $CsH_2PO_4$ has been found to be a promising candidate for $H_2/O_2$ and DMFC systems. It undergoes a superprotonic phase transition at 230 °C [94, 95]. However, in reducing and oxidising environments, $CsH_2PO_4$ also undergoes dehydration, but a toxic gas is not formed in this case. Other group-IVB metals such as titanium and zirconium are potential candidates but studies on them have been sparse. Monoclinic phases in the form $MH(PO_3H)$, where M can be sodium, potassium, rubidium, caesium or $NH_4$, have also been investigated for superprotonic behaviour [96]. Proton conduction in such systems takes place via the Grotthuss mechanism with the help of two processes: (i) by movement of protons between adjacent oxy-anion tetrahedrons and (ii) by splitting of O–H–O bonds and reorientation of oxy-anion groups [97]. However, studies have reported the use of solid acids as composites with other polymers due to the poor mechanical properties and low ductility, making formation of thin membranes difficult.

### 2.4.2 Carbon nanomaterials

Carbon nanomaterials have been very popular in fuel cells for use in electrodes, GDL and membranes. Carbon nanomaterials such as multi-walled CNT or multi-walled CNT along with GO and their various chemically-modified variants have been formed into membranes and tested in PEMFC and DMFC systems. While carbon nanofibres have been woven into GDLs, CNT along with their chemically-modified versions (sulfonated, carboxylated, phosphonated) have been explored mostly as fillers in composite membranes. Studies have explored the effect of incorporating CNT and functionalised CNT as filler for Nafion® and PBI membranes [50–52]. GO has been formed into free-standing films (formed mostly using filtration), which have been used in multilayer and single-layer membranes [98–101]. CNT can also be formed into films using filtration or casting, but their application as membranes has been very limited. A recent study incorporated a sulfonated inner layer of CNT with outer layers of sulfonated polyarylene ether nitrile to form a three-layered membrane for application in DMFC that revealed higher proton conductivity (0.275 S/cm) and lower MeOH permeation than Nafion® 117 at 80 °C [102]. These membranes were prepared using solution casting and the resultant multilayers also demonstrated high mechanical strength (42.35 MPa) in wet state due to incorporation of CNTs.

# References

[1]  *Fuel Cell Technologies Office Multi-Year Research, Development, and Demonstration Plan*, US Department of Energy, Washington, DC, USA, 2017. https://energy.gov/eere/fuelcells/down loads/fuelcell-technologies-office-multi-year-research-development-and-22 [last accessed May 2017].

[2]  *Fuel Cell System Cost – 2016*, US Department of Energy, Washington, DC, USA, 2016. https://www.hydrogen.energy.gov/pdfs/16020_fuel_cell_system_cost_2016.pdf [last accessed May 2017].

[3]  F. Barbir in *PEM Fuel Cells: Theory and Practice*, Academic Press, Cambridge, MA, USA, 2012, p.76.

[4]  S. Gottesfeld and T.A. Zawodzinski in *Advances in Elelctrochemical* Science *and* Engineering, Volume 5, Eds., R.C. Alkire, H. Gerischer, D.M. Kolb and C.W. Tobias, Wiley-VCH, New York, NY, USA, 1997, p.195.

[5]  R. O'hayre, S. Cha, F.B. Prinz and W. Colella in *Fuel Cell Fundamentals*, Anonymous R. John Wiley & Sons, New York, NY, USA, 2009, p.1.

[6]  V. Rao, N. Kluy, W. Ju and U. Stimming in *Handbook of Membrane Separations: Chemical, Pharmaceutical, Food, and Biotechnological Applications*, Eds., A.K. Pabby, S.S. Rizvi and A.M.S. Requena, CRC Press, Boca Raton, FL, USA, 2015, p.567.

[7]  A.B. LaConti, A.R. Fragala and J.R. Boyak in *Proceedings of the Symposium on Electrode Materials and Processes for Energy Conversion and Storage*, Eds., J.D.E. McIntyre, S. Srinivasan and F.G. Will, The Electrochemical Society, Princeton, NJ, USA, 1977, p.354.

[8]  T.A. Zawodzinski, T.E. Springer, J. Davey, R. Jestel, C. Lopez, J. Valerio and S. Gottesfeld, *Journal of the Electrochemical Society*, 1993, **140**, 7, 1981.

[9]  T.F. Fuller and J. Newman, *Journal of the Electrochemical Society*, 1992, **139**, 5, 1332.

[10]  T.A. Zawodzinski, J. Davey, J. Valerio and S. Gottesfeld, *Electrochimica Acta*, 1995, **40**, 3, 297.

[11]  Q. Yan, H. Toghiani and J. Wu, *Journal of Power Sources*, 2006, **158**, 1, 316.

[12]  W. Dai, H. Wang, X. Yuan, J.J. Martin, D. Yang, J. Qiao and J. Ma, *International Journal of Hydrogen Energy*, 2009, **34**, 23, 9461.

[13]  F.N. Büchi and G.G. Scherer, *Journal of the Electrochemical Society*, 2001, **148**, 3, A183.

[14]  G. Janssen and M. Overvelde, *Journal of Power Sources*, 2001, **101**, 1, 117.

[15]  A. Iranzo and P. Boillat, *International Journal of Hydrogen Energy*, 2014, **39**, 30, 17240.

[16]  N. Radenahmad, A. Afif, P.I. Petra, S.M.H. Rahman, S.G. Eriksson and A.K. Azad, *Renewable and Sustainable Energy Reviews*, 2016, **57**, 1347.

[17]  A. Kraytsberg and Y. Ein-Eli, *Energy & Fuels*, 2014, **28**, 12, 7303.

[18]  N. Awang, A.F. Ismail, J. Jaafar, T. Matsuura, H. Junoh, M.H.D. Othman and M.A. Rahman, *Reactive and Functional Polymers*, 2015, **86**, 248.

[19]  S.J. Peighambardoust, S. Rowshanzamir and M. Amjadi in *Review of the Proton Exchange Membranes for Fuel Cell Applications*, Elsevier Ltd, Amsterdam, The Netherlands, 2010, p.9349.

[20]  T. Kinumoto, M. Inaba, Y. Nakayama, K. Ogata, R. Umebayashi, A. Tasaka, Y. Iriyama, T. Abe and Z. Ogumi, *Journal of Power Sources*, 2006, **158**, 2, 1222.

[21]  K.D. Kreuer, *Journal of Membrane Science*, 2001, **185**, 1, 29.

[22]  J. Huslage, T. Rager, B. Schnyder and A. Tsukada, *Electrochimica Acta*, 2002, **48**, 3, 247.

[23]  C. Stone, A.E. Steck and R.D. Lousenberg, inventors; Ballard Power Systems Inc., assignee; US5602185A, 1997.

[24]  M. Sankir, Y.S. Kim, B.S. Pivovar and J.E. McGrath, *Journal of Membrane Science*, 2007, **299**, 1, 8.

[25]   B. Campagne, G. David, B. Améduri, D.J. Jones, J. Rozière and I. Roche, *Macromolecules*, 2013, **46**, 8, 3046.

[26]   Q. Li, R. He, J.O. Jensen and N.J. Bjerrum, *Chemistry of Materials*, 2003, **15**, 26, 4896.

[27]   M. Rikukawa and K. Sanui, *Progress in Polymer Science (Oxford)*, 2000, **25**, 10, 1463.

[28]   D. Wu, T. Xu, L. Wu and Y. Wu, *Journal of Power Sources*, 2009, **186**, 2, 286.

[29]   R. He, Q. Li, G. Xiao and N.J. Bjerrum, *Journal of Membrane Science*, 2003, **226**, 1, 169.

[30]   M. Hazarika and T. Jana, *ACS Applied Materials & Interfaces*, 2012, **4**, 10, 5256.

[31]   S. Wang, C. Zhao, W. Ma, G. Zhang, Z. Liu, J. Ni, M. Li, N. Zhang and H. Na, *Journal of Membrane Science*, 2012, **411**, 54.

[32]   C. Shen, L. Jheng, S.L. Hsu and J.T. Wang, *Journal of Materials Chemistry*, 2011, **21**, 39, 15660.

[33]   Q. Tang, G. Qian and K. Huang, *RSC Advances*, 2013, **3**, 11, 3520.

[34]   Z. Guo, R. Xiu, S. Lu, X. Xu, S. Yang and Y. Xiang, *Journal of Materials Chemistry A*, 2015, **3**, 16, 8847.

[35]   P.K. Varshney and S. Gupta, *Ionics*, 2011, **17**, 6, 479.

[36]   H. Vaghari, H. Jafarizadeh-Malmiri, A. Berenjian and N. Anarjan, *Sustainable Chemical Processes*, 2013, **1**, 1, 16.

[37]   Y. Ye, J. Rick and B. Hwang, *Polymers*, 2012, **4**, 4, 913.

[38]   C.W. Lin, S.S. Liang, S.W. Chen and J.T. Lai, *Journal of Power Sources*, 2013, **232**, 297.

[39]   M. Santamaria, C. Pecoraro, F. Di Franco, F. Di Quarto, I. Gatto and A. Saccà, *International Journal of Hydrogen Energy*, 2016, **41**, 11, 5389.

[40]   S.P. Cabello, N. Ochoa, E. Takara, S. Mollá and V. Compañ, *Carbohydrate Polymers*, 2017, **157**, 1759.

[41]   B.P. Tripathi and V.K. Shahi, *The Journal of Physical Chemistry B*, 2008, **112**, 49, 15678.

[42]   B. Smitha, S. Sridhar and A. Khan, *Journal of Power Sources*, 2006, **159**, 2, 846.

[43]   S. Mohanapriya, S. Bhat, A. Sahu, A. Manokaran, R. Vijayakumar, S. Pitchumani, P. Sridhar and A. Shukla, *Energy & Environmental Science*, 2010, **3**, 11, 1746.

[44]   Y. Xiao, Y. Xiang, R. Xiu and S. Lu, *Carbohydrate Polymers*, 2013, **98**, 1, 233.

[45]   S. Mohanapriya, G. Rambabu, S. Suganthi, S. Bhat, V. Vasanthkumar, V. Anbarasu and V. Raj, *RSC Advances*, 2016, **6**, 62, 57709.

[46]   N. Shaari and S.K. Kamarudin, *Malaysian Journal of Analytical Sciences*, 2017, **21**, 1, 113.

[47]   H. Nagar, C. Sumana, V.V.B. Rao and S. Sridhar, *Journal of Applied Polymer Science*, 2017, **134**, 7, 1.

[48]   C.M. Branco, S. Sharma, M.M. de Camargo Forte and R. Steinberger-Wilckens, *Journal of Power Sources*, 2016, **316**, 319.

[49]   K. Ketpang, K. Lee and S. Shanmugam, *ACS Applied Materials & Interfaces*, 2014, **6**, 19, 16734.

[50]   Y. Liu, B. Yi, Z. Shao, D. Xing and H. Zhang, *Electrochemical and Solid-State Letters*, 2006, **9**, 7, A356.

[51]   R. Kannan, B.A. Kakade and V.K. Pillai, *Angewandte Chemie International Edition*, 2008, **47**, 14, 2653.

[52]   R. Kannan, H.N. Kagalwala, H.D. Chaudhari, U.K. Kharul, S. Kurungot and V.K. Pillai, *Journal of Materials Chemistry*, 2011, **21**, 20, 7223.

[53]   Y. Kim, K. Ketpang, S. Jaritphun, J.S. Park and S. Shanmugam, *Journal of Materials Chemistry A*, 2015, **3**, 15, 8148.

[54]   A. Shukla, S.D. Bhat and V.K. Pillai, *Journal of Membrane Science*, 2016, **520**, 657.

[55]   A. Ranade, N.A. D'Souza and B. Gnade, *Polymer*, 2002, **43**, 13, 3759.

[56]   J. Chang, J.H. Park, G. Park, C. Kim and O.O. Park, *Journal of Power Sources*, 2003, **124**, 1, 18.

[57]   Y. Kojima, K. Fukumori, A. Usuki, A. Okada and T. Kurauchi, *Journal of Materials Science Letters*, 1993, **12**, 12, 889.

[58]  C.H. Rhee, H.K. Kim, H. Chang and J.S. Lee, *Chemistry of Materials*, 2005, **17**, 7, 1691.

[59]  M.M. Hasani-Sadrabadi, S.H. Emami, R. Ghaffarian and H. Moaddel, *Energy & Fuels*, 2008, **22**, 4, 2539.

[60]  S.K. Nataraj, C. Wang, H. Huang, H. Du, L. Chen and K. Chen, *ACS Sustainable Chemistry & Engineering*, 2015, **3**, 2, 302.

[61]  M. Purwanto, L. Atmaja, M.A. Mohamed, M.T. Salleh, J. Jaafar, A.F. Ismail, M. Santoso and N. Widiastuti, *RSC Advances*, 2016, **6**, 3, 2314.

[62]  I.G.B.N. Makertihartha, M. Zunita, Z. Rizki and P.T. Dharmawijaya in *Proceedings AIP Conference 2017*, AIP Publishing LLC, 2017, **1818**, 1, 020030–10.

[63]  K.L. Yeung and W. Han, *Catalysis Today*, 2014, **236**, Part B, 182.

[64]  J. Wang, X. Zheng, H. Wu, B. Zheng, Z. Jiang, X. Hao and B. Wang, *Journal of Power Sources*, 2008, **178**, 1, 9.

[65]  W. Han, S.M. Kwan and K.L. Yeung, *Topics in Catalysis*, 2010, **53**, 19–20, 1394.

[66]  S.P. Jiang, *Journal of Materials Chemistry A*, 2014, **2**, 21, 7637.

[67]  Z. Zhang, F. Désilets, V. Felice, B. Mecheri, S. Licoccia and A.C. Tavares, *Journal of Power Sources*, 2011, **196**, 22, 9176.

[68]  C. Beauger, G. Lainé, A. Burr, A. Taguet, B. Otazaghine and A. Rigacci, *Journal of Membrane Science*, 2013, **430**, 167.

[69]  B.P. Tripathi, M. Kumar and V.K. Shahi, *Journal of Membrane Science*, 2009, **327**, 1, 145.

[70]  E.N. Gribov, E.V. Parkhomchuk, I.M. Krivobokov, J.A. Darr and A.G. Okunev, *Journal of Membrane Science*, 2007, **297**, 1, 1.

[71]  B. Libby, W. Smyrl and E. Cussler, *AIChE Journal*, 2003, **49**, 4, 991.

[72]  X. Li, E.P. Roberts, S.M. Holmes and V. Zholobenko, *Solid State Ionics*, 2007, **178**, 19, 1248.

[73]  B.A. Holmberg, X. Wang and Y. Yan, *Journal of Membrane Science*, 2008, **320**, 1, 86.

[74]  Y. Wang, D. Yang, X. Zheng, Z. Jiang and J. Li, *Journal of Power Sources*, 2008, **183**, 2, 454.

[75]  F.G. Üçtu and S.M. Holmes, *Electrochimica Acta*, 2011, **56**, 24, 8446.

[76]  E. Bakangura, L. Wu, L. Ge, Z. Yang and T. Xu, *Progress in Polymer Science*, 2016, **57**, 103.

[77]  J.A. Hurd, R. Vaidhyanathan, V. Thangadurai, C.I. Ratcliffe, I.L. Moudrakovski and G.K.H. Shimizu, *Nature Chemistry*, 2009, **1**, 9, 705.

[78]  M. Díaz, A. Ortiz, M. Vilas, E. Tojo and I. Ortiz, *International Journal of Hydrogen Energy*, 2014, **39**, 8, 3970.

[79]  M. Armand, F. Endres, D.R. MacFarlane, H. Ohno and B. Scrosati, *Nature Materials*, 2009, **8**, 8, 621.

[80]  A. Fernicola, S. Panero, B. Scrosati, M. Tamada and H. Ohno, *ChemPhysChem*, 2007, **8**, 7, 1103.

[81]  F. Chu, B. Lin, F. Yan, L. Qiu and J. Lu, *Journal of Power Sources*, 2011, **196**, 19, 7979.

[82]  A.N. Mondal, B.P. Tripathi and V.K. Shahi, *Journal of Materials Chemistry*, 2011, **21**, 12, 4117.

[83]  R.S. Malik, S.N. Tripathi, D. Gupta and V. Choudhary, *International Journal of Hydrogen Energy*, 2014, **39**, 24, 12826.

[84]  J. Yang, Q. Che, L. Zhou, R. He and R.F. Savinell, *Electrochimica Acta*, 2011, **56**, 17, 5940.

[85]  S. Sekhon, P. Krishnan, B. Singh, K. Yamada and C. Kim, *Electrochimica Acta*, 2006, **52**, 4, 1639.

[86]  J. Sun, L. Jordan, M. Forsyth and D. MacFarlane, *Electrochimica Acta*, 2001, **46**, 10, 1703.

[87]  Y. Ye, C. Tseng, W. Shen, J. Wang, K. Chen, M. Cheng, J. Rick, Y.Huang, F. Chang and B. Hwang, *Journal of Materials Chemistry*, 2011, **21**, 28, 10448.

[88]  C. Xu, X. Liu, J. Cheng and K. Scott, *Journal of Power Sources*, 2015, **274**, 922.

[89]  L.A. Neves, I.M. Coelhoso and J.G. Crespo, *Journal of Membrane Science*, 2010, **360**, 1, 363.

[90]  G. Lakshminarayana and M. Nogami, *Electrochimica Acta*, 2010, **55**, 3, 1160.

[91]  W. Li, F. Zhang, S. Yi, C. Huang, H. Zhang and M. Pan, *International Journal of Hydrogen Energy*, 2012, **37**, 1, 748.

[92]  P. Kiatkittikul, T. Nohira and R. Hagiwara, *Journal of Power Sources*, 2012, **220**, 10.

[93]  A. Ortiz-Negrón, N. Lasanta-Cotto and D. Suleiman, *Journal of Applied Polymer Science*, 2017, **134**, 22,

[94]  D.A. Boysen, T. Uda, C.R. Chisholm and S.M. Haile, *Science*, 2004, **303**, 5654, 68.

[95]  S. Singhal, *Solid State Ionics*, 2000, **135**, 1, 305.

[96]  W. Zhou, A.S. Bondarenko, B.A. Boukamp and H.J. Bouwmeester, *Solid State Ionics*, 2008, **179**, 11, 380.

[97]  K.D. Kreuer, *Solid State Ionics*, 1997, **94**, 55.

[98]  K. Ravikumar and K. Scott, *Chemical Communications*, 2012, **48**, 45, 5584.

[99]  T. Bayer, S.R. Bishop, M. Nishihara, K. Sasaki and S.M. Lyth, *Journal of Power Sources*, 2014, **272**, 239.

[100]  C.W. Lin and Y.S. Lu, *Journal of Power Sources*, 2013, **237**, 187.

[101]  Z. Jiang, Y. Shi, Z. Jiang, X. Tian, L. Luo and W. Chen, *Journal of Materials Chemistry A*, 2014, **2**, 18, 6494.

[102]  M. Feng, Y. You, P. Zheng, J. Liu, K. Jia, Y. Huang and X. Liu, *International Journal of Hydrogen Energy*, 2016, **41**, 9,5113.

# 3 Proton-conducting membranes: Preparation methods

## 3.1 Introduction

Due to the wide range of membrane applications, membrane- preparation methods have been studied extensively since the early-1900s [1]. Over time, a vast knowledge base has been accrued with regard to the mechanisms and thermodynamics of polymer-membrane formation. This has led to in-depth understanding of the various aspects of several preparation methods for application in ion separation, water purification/desalination, and dialysis. Many of these methods have been adopted in the preparation of membranes aimed at fuel cell applications.

The preparation method can have an enormous impact on the properties of the prepared membrane. The type of polymer/material and required structure of the membrane based on its intended application, in general, govern the choice of the preparation method and the parameters involved therein. Most commercially available membranes made of perfluorosulfonic acid are usually extruded and solution-cast films of different thickness. Many commercial companies also offer aqueous/alcohol dispersions and pellets. Aqueous dispersions are often utilised in the laboratory to prepare solution-cast Nafion® and Nafion® composites for experimental purposes. Extrusion has been a well-developed polymer processing technology since the 1930s and requires specialised equipment. The process can be used to produce fibres as well as polymer films using monomers (*via* reactive extrusion) as well as polymers (*via* melt extrusion). In the absence of detailed accounts of the synthesis and processing of Nafion® membranes, in general it is considered that the copolymer is melt-extruded in a sulfonyl fluoride precursor to form a membrane with sulfonic acid functionality [2]. Extruded Nafion® membranes possess better proton conductivity under high-temperature low-humidity conditions as compared with solution-cast Nafion® membranes. Most commercial companies making membranes also promote reinforced membranes (such as Nafion® XL, Aciplex™-S-AH and Aciplex™-S-H-EH), which are essentially strengthened (or reinforced) using polytetrafluoroethylene (PTFE) fibres, which facilities thinner membranes without compromising on mechanical strength. Reinforced and chemically-stabilised membranes also demonstrate longer operating durability and tolerance to degradation as opposed to cast membranes. Different approaches are used for reinforcement, as shown below:

- Microporous-stretched PTFE filled with a perfluorosulfonic ionomer (as in the case of PRIMERA® by Gore® Select).
- Dispersion of a small amount of PTFE fibres in a polymer matrix of perfluorosulfonic resin with high ion-exchange capacity followed by extrusion and uniform stretching and further treatments with acid and alkali (as in case of Flemion™ by AGC manufacturers) [3, 4].

https://doi.org/10.1515/9783110647327-003

Reports on extrusion-based membrane synthesis are sparse due to limited access to specialised equipment. Thomassin and co-workers reported the incorporation of multi-walled carbon nanotubes (MWCNT) and carboxylated MWCNT in Nafion® using melt extrusion to reveal reduced methanol permeability without negatively impacting the ionic conductivity of the membranes [5].

Apart from extrusion, one of the most widely utilised and oldest concept for membrane preparation is phase inversion (i.e., the process of transforming a polymer from a solution phase to a solid phase in a controlled manner). This process underlies many membrane-preparation methods, such as evaporation-induced phase separation. Electrospinning and dip coating are also reported in literature regularly for preparation of novel membranes. This chapter focuses on the methods reported recently for membranes intended for fuel cell application.

## 3.2 Solution casting

Solution casting (also known as evaporation-induced phase separation) is the most widely used, facile method for preparation of a small number of membrane samples, especially on a laboratory scale. There is more than one variation of this method. In general, a viscous polymer solution is first prepared in a solvent (this can also be a mixture of two or more solvents). Sometimes the solvent mixture may also include a non-solvent. The prepared solution is then cast evenly in the form of a film on a flat substrate using a 'casting knife' (also known as a 'doctor blade'). Figure 3.1 shows a

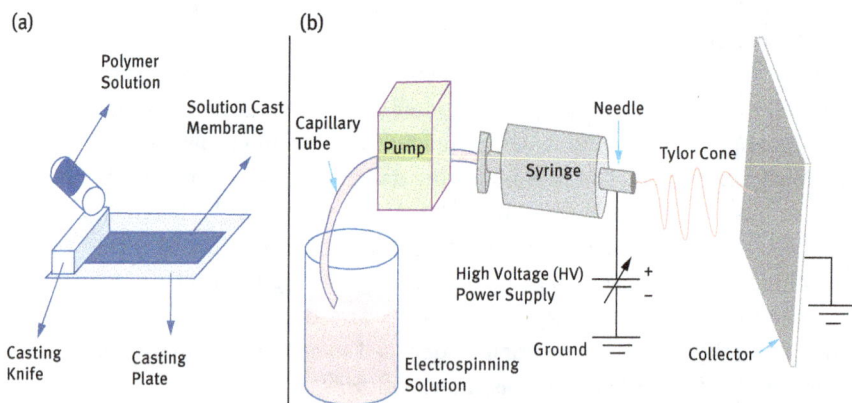

**Figure 3.1:** Schematics showing (a) solution casting. Reproduced with permission from V. Salles, Laurence Seveyrat, T. Fiorido, L. Hu, J. Galineau, C. Eid, B. Guiffard, A. Brioude and D. Guyomar in *Nanowires – Recent Advances*, Ed., X. Peng, InTech, Rijeka, Croatia, 2012, p.295. ©2012, InTech [13]. A homemade setup for electrospinning preparation processes. Reproduced with permission from D. Hassan, S. El-safty, K.A. Khalil, M. Dewidar and G. Abu El-magd, *Materials*, 2016, **9**, 9, 759. ©2016, MDPI [14].

simple schematic. The volatile solvent then evaporates to leave behind a polymer film. Another laboratory variation of this method involves the pouring of the solution slowly (to prevent formation of air bubbles) into a petri dish, which is then allowed to dry. The volume of the solution poured into the dish depends of the concentration of polymer solution and the thickness of the final membrane. Factors that impact the structure and performance of the as-prepared membrane for a given polymer during this process include:

– Type of solvent (its boiling point and volatility)
– Composition of the casting solution
– Crystallisation, vitrification and gelation behaviour of the polymer
– Duration of solvent evaporation
– Temperature

Variation in any of these parameters will have an effect on the final structure, porosity and mechanical strength of the membrane [6]. Moderately volatile solvents are usually preferred because rapid evaporation (in the case of highly volatile solvents) cools the solvent and results in gelation of the polymer, leading to a 'corrugated' or 'mottled' structure. Choice of the solvent and the concentration of the polymer also play important part in defining the crystallinity of the membrane [7]. Crystallisation occurs if the solution temperature is lower than the melting point of the polymer. However, the polymer chains can form lamellar-type or dendrite-like complex morphologies as they crystallise depending upon whether the solution is very dilute or moderately concentrated. Semi-crystalline polymers can form membranes with very little or no crystalline content. If the solution temperature becomes higher than the glass transition temperature ($T_g$) of the polymer, this leads to vitrification (i.e., formation of a glassy, brittle, amorphous solid) [6].

This method is used commonly in laboratory studies to cast Nafion® and other polymeric as well as composite membranes [8, 9]. Silva and co-workers explored the role of different solvents in cast Nafion® membranes and their properties [10]. Solution-cast membranes are commonly subjected to further annealing at higher temperatures (just below the $T_g$) to improve their mechanical strength *via* further crosslinking of polymer chains. Studies have demonstrated an increase in the water-uptake capacity of the cast Nafion® upon annealing [11, 12].

This method has also been explored for the preparation of multilayered membranes. In this case, once the solvent has evaporated and a layer of polymer has been cast, another solution of a different polymer is then poured into the petri dish and dried in the same way. This process can be continued depending upon the number of layers desired [15] and the final thickness of the multilayer membrane. Once the layering is complete, the final layered structure is further heat-treated to allow inter- and intra-layer crosslinking. The primary considerations in the case of a 'N'-numbered multilayer is the compatibility of adjacent layers (to ensure good adhesion and interlayer interaction for proton conduction), choice of solvents (to ensure the formation of one layer does not interfere with other layers). This concept has attracted attention for the preparation of direct methanol fuel cell (DMFC) and proton-exchange membrane fuel cell (PEMFC) membranes [16–21]. Hybrid multilayers comprising outer layers made of polymers and inner composite layers (polymer layers with fillers such as silicon dioxide) have also been explored [22–24]. Another example of a multilayer is the use of barrier coatings for existing polymer membranes. Shao and co-workers demonstrated a protective barrier coating for Nafion® prepared by solution casting polyvinyl alcohol (PVA) to reduce methanol crossover [25]. Solution casting is one of the more successful methods of multilayer preparation because it allows a significant degree of chemical and mechanical interaction between adjacent layers, thereby reducing the risk of delamination while promoting proton conductivity.

## 3.3 Electrospinning

Electrospinning is a versatile method for producing nanofibres. Although used for more than 70 years, the process is relatively new for applications such as fuel cell and other porous membrane (for filtration and desalination) fabrications. Electrospinning involves application of a high-potential electrostatic field between a grounded collector and a droplet of polymer solution. As the electrostatic potential increases, it overcomes the surface-tension forces of the droplet. This induces the formation of a charged liquid polymer jet leaving the spinneret. As the surface tension decreases, the droplet forms a cone (called a 'Taylor cone') before eventually escaping the spinneret tip and forming into a charged jet [7, 26]. The route of the charged liquid jet is controlled by an electric field. Consequently, the liquid jet forms into a fibre and the final solidified spun fibre is collected onto a stationary or rotating-conductive collector, as shown in Figure 3.1. This method offers advantages such as control over the fibre aspect ratio (ratio of length:diameter) and fibre morphology. By varying parameters such as the viscosity of the polymer solution, applied potential, and flow rate of the solution to alter the properties mentioned above, the pore size, porosity and hydrophobicity can be 'tailored'. Table 3.1 lists the effect of variation in electrospinning parameters on the proton conductivity and methanol permeability of fuel cell membranes. This process can also be used for composite membranes.

Studies on Nafion® and other polymer (sulfonated polyimide) membranes prepared using electrospinning provide evidence that better control over fibre diameter (allowing thinner fibres) along with the orientation of the ionic domains along the fibre axis (achieved due to shear force during electrospinning) enables significant enhancement in proton conductivity and, in some cases, reduced gas crossover attributed to an aggregated structure within the nanofibres [26, 28, 29]. A study by Dong and co-workers revealed that the electrospun nanofibre (diameter 400 nm) of

**Table 3.1:** Effect of variation in electrospinning parameters on the proton conductivity and methanol permeability of fuel cell membranes.

| Parameter | Affect on fibre structure | Effect on proton conductivity | Effect on methanol permeability* |
|---|---|---|---|
| Increase in applied voltage | Fibre diameter decreases | Increases | Decreases |
| Increase in needle to collector distance | Fibre diameter decreases | Increases | Decreases |
| Increase in viscosity of dope solution | Smooth fibre (without beads) | Increases | Decreases |

* The effect on methanol permeability is also contributed by the filler within the parent polymer matrices

Reproduced with permission from H. Junoh, J. Jaafar, M. Noorul, A. Mohd, A.F. Ismail, M. Hafiz, D. Othman, M.A. Rahman, N. Yusof, W. Norhayati, W. Salleh and H. Ilbeygi, *Journal of Nanomaterials*, 2014,**2015**, 1. ©2014, Hindawi [27] and H. Junoh, J. Jaafar, M.N.A.M. Norddin, A.F. Ismail, M.H.D. Othman, M.A. Rahman, N. Yusof, W.N.W. Salleh andH. Ilbeygi, *Journal of Nanomaterials*, 2015, **690965**, 1. ©2015, Hindawi [28]

highly-purity Nafion® (prepared using 0.1 wt% of high-molecular-weight polyethylene oxide as the carrier polymer) exhibited proton conductivity (1.5 S/cm) one order of magnitude higher as compared with that of bulk Nafion® (0.1 S/cm). The fibre also demonstrated one order-higher humidity sensitivity [26] Table 3.2 details selected reports on electrospun fuel cell membranes and their preparation parameters. Other studies have also looked at the effects of electrospinning on polymer side chains and use of dual polymers for electrospun-mat structures for fuel cells [30]. However, more studies are required to achieve better control and repeatability over electrospinning.

## 3.4 Multilayer membrane systems

The concept of multilayer membranes entails the idea of combining the best properties of polymers as well as other non-polymeric materials without having to compromise on other aspects. This concept has been known for quite a while in the membrane industry for applications in nanofiltration, dialysis, and ion separation, but is fairly novel for research on fuel cell membranes [15, 35–41]. Multilayers can be made from any number of layers of required thicknesses with desired combinations of materials. The concept is likely to give rise to a multitude of permutations given the variety of materials used in fuel cell membranes. A schematic showing the generic preparation of a multilayer membrane is shown in Figure 3.2. In recent years, multilayers have been studied particularly for DMFC, PEMFC and alkali fuel cell applications [23, 24, 42–44]. Some of the specific motivations for multilayer

**Table 3.2:** Details of electrospinning parameters used and performance observations for polymer membranes used in fuel cells.

| Polymer | Type of fuel cell | Electrospinning parameters | Remark | Ref. |
|---|---|---|---|---|
| PVA/Nafion® | DMFC | Distance from needle to collector: 20 cm<br>Solution flow rate: 1.2 ml/h<br>Voltage potential: 20 kV<br>I.D. needle: 0.8 mm<br>Collector: Cu collection roll | More straight and less tortuous effect on Nafion®/PVA nanofibre which leads to an increase in proton conductivity of the membrane and reduction in membrane crossover with a thickness of ≈50 μm | [32] |
| SiO$_2$/SPEEK/ Nafion® | PEMFC | Distance from needle to collector: 10 cm<br>Voltage potential: 15 kV<br>I.D. needle: 0.15 mm | Thickness of SiO$_2$/SPEEK nanofibre membrane 45-μm incorporated with Nafion® produces high-proton conductivity compared with cast Nafion® and SPEEK. The maximum power density of SiO$_2$/SPEEK incorporation with Nafion® is 170 mW/cm3, double that of cast Nafion® | [33] |
| PVA/Nafion® | DMFC | Distance from needle to collector: 25 cm<br>Solution flow rate: 0.5 ml/h<br>Voltage potential: 16 kV | Succeeds in producing PVA nanofibres having a diameter of 200–300 nm. Contribution of Nafion® as the support material increases the mechanical and thermal properties of the composite membrane, with a thickness of the composite membrane of 46–47μm. PVA nanofibres have increased the barrier properties on methanol crossover. | [34] |
| Sulfonated polyether sulfone, SPES/ Nafion® | DMFC | Distance from needle to collector: 15 cm<br>Voltage potential: 19 kV<br>Collector: rotating cylindrical stainless steel | Proton conductivity of the bilayer SPES/Nafion® barely changes compared with that of Nafion® 117/112. The methanol crossover has decreased due to the SPES nanofibres within Nafion® matrices | [35] |

SPEEK: Sulfonated polyether ether ketone
SPES: Sulfonated polyether sulfone

Reproduced with permission from H. Junoh, J. Jaafar, M. Noorul, A. Mohd, A.F. Ismail, M. Hafiz, D. Othman, M.A. Rahman, N. Yusof, W. Norhayati, W. Salleh and H. Ilbeygi, *Journal of Nanomaterials*, 2014, **2015**, 1. ©2014, Hindawi [27] and H. Junoh, J. Jaafar, M.N.A.M. Norddin, A.F. Ismail, M.H.D. Othman, M.A. Rahman, N. Yusof, W.N.W. Salleh and H. Ilbeygi, *Journal of Nanomaterials*, 2015, **690965**, 1. ©2015, Hindawi [28]

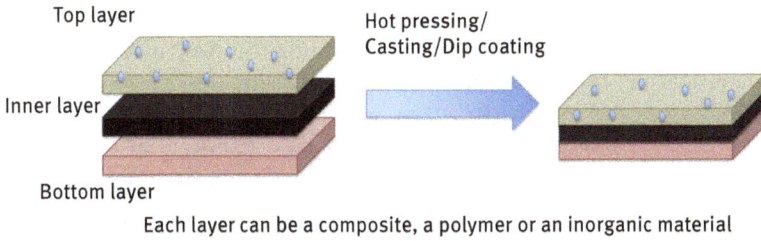

Each layer can be a composite, a polymer or an inorganic material

**Figure 3.2:** Basic concept of a multilayer membrane (schematic). Reproduced with permission from C.M. Branco, S. Sharma, M.M. De Camargo Forte and R. Steinberger-Wilckens, *Journal of Power Sources*, 2016, **316**, 139. ©2016, Elsevier [15].

membranes are developing membranes capable of 'self-humidification'. These include highly sulfonated materials for long-term retention of water to operate in low-humidity conditions and high-methanol tolerance without compromising on mechanical strength, durability and range of operating temperatures. Table 3.3 compares the pros and cons of composite and multilayer membranes.

**Table 3.3:** Comparison of the two approaches to membrane preparation and novel trends specific to them.

| Approach/ method | Composite membranes | Multilayer membranes |
|---|---|---|
| Advantages | Fast manufacture Known physical and chemical proprieties | Countless combination of materials and layers Keeps the characteristics of each layer intact |
| Disadvantages | Difficult homogeneity Polymer matrix must be resistant to water | Extra interface problematic Longer time of manufacture |
| New trend | Use of carbon nanostructures and ionic liquids | Inner layer not just as a polymer, but with a complex formulation |

Reproduced with permission from C.M. Branco, S. Sharma, M.M. De Camargo Forte and R. Steinberger-Wilckens, *Journal of Power Sources*, 2016, **316**, 139. ©2016, Elsevier [15]

With the development of multilayer membranes, other innovative membrane-preparation methods have been explored. Although not entirely novel, these methods in many cases are new to membrane processing for fuel cell membranes. Some of these are detailed in this section.

## 3.4.1 Dip coating

Dip coating is a commonly used method for developing coatings for various applications. With its industrial use going back to the 1940s, it is used widely for coating glass and other optical coatings on an industrial scale. It is also extremely popular for laboratory-scale processing due to its simple setup and low cost [45]. Polymeric and metallic materials can be dip coated by dipping them into desired solutions (of different metal/non-metallic nanoparticles, or polymers) followed by their withdrawal from the solution. Depending on the requirement, the cycle of immersion and withdrawal can be repeated a number of times. Annealing or some kind of curing usually follows to consolidate the coating thus formed. When incorporating this principle in the preparation of composite or multilayer membranes, it is essential to have a starting layer with good mechanical strength. The process of dip coating is also referred to as a 'layer-by-layer' assembly process. Various reports of multilayer membranes have surfaced in recent years, which have utilised this method for DMFC and PEMFC application [46–49]. Table 3.4 lists some of these recent reports. This process allows the coating of ultra-thin (≤1–2 μm) layers and, in principle, can be used to coat as many bilayers once a previously coated layer is sufficiently dry. Parameters such as the hydrophility/hydrophobicity of the starting layer, choice of solvent for coating solution, coating-solution concentration, solution temperature, and dipping speed/rate can be explored to optimise the properties of the layer to be coated.

### 3.4.2 Hot pressing

Hot pressing is utilised regularly for preparing membrane electrode assembly (MEA) for fuel cells. The membrane is placed between the gas diffusion layers (GDL) coated with electrocatalyst ink and hot pressed together. A similar process can be used when combining two or more membranes to form a multilayer-membrane system. Single layers of polymer (produced by solution casting, commercial extrusion, or electrospinning) or non-polymer or even composites can be combined into a multilayer membrane by hot pressing [15, 42–44, 59]. High temperature and pressure together enable the formation of mechanical bonds between the adjacent layers. This is one of the simplest approaches adopted for multilayer-membrane preparation [15]. When using hot pressing, the temperature, pressure, and compatibility of all polymer/non-polymer layers with these two vital parameters must be considered very carefully.

### 3.4.3 Filtration

Vacuum-assisted filtration is used for preparing free-standing graphene oxide (GO)-based thin films to be used independently or in multilayer membranes [60–63]. Membranes prepared from chemically-modified carbon nanotubes and GO have also been studied for water purification and gas separation due to their antifouling property and nanofiltration capabilities [64, 65]. GO-based membranes prepared using

**Table 3.4:** Properties of multilayer membranes prepared using dip coating reported in the literature.

| Fuel cell | Layer materials | Number of layers | σ (S/cm) | Water uptake (%) | Ref |
|---|---|---|---|---|---|
| PEMFC | N/sulfonated polyimide/N | 3 | 0.07 (at 80 °C) | NA | [47] |
| | SPEEK + PDDA/PSS | N/A | 0.061 (at 80 °C, 100% RH) | 47 (at 30 °C) | [51] |
| DMFC | N + sulfonated cardo polyarylene ether sulfone/ glutaraldehyde | 1 + 1 to 50 × 2 (dip both sides) | 0.067 (for 25 × 2 layers, at 30 °C) | 21.3 | [52] |
| | SPAEK + polyaniline | 1 + (5 × 2) | 0.24 (at 80 °C) | 93.8 | [53] |
| | N + PAH/PSS + salt | 1 + 5 up to 20 × 2 | 0.08791 (for 5 × 2 layers, in 0.1 M NaCl, 22 °C) | NA | [54] |
| | N + heteropolytungstate PDDA | 1 + 1 up to 5 × 2 | 0.03 (5 × 2 layers) | NA | [55] |
| | Polyethylene/PSS + polyvinylimidazole/ polyacrylamide methyl propane sulfonic acid | 1 + (1 × 2) | 0.122 | NA | [56] |
| | SPAEK + phosphotungstic acid/PPy | 1 + 1 to 5 × 2 | 0.299 (for 5 × 2 layers, 80 °C) | 49.85 | [57] |
| | N + sulfonated polyaryl ether ketone bearing carboxyl groups/chitosan | N/A | 0.131 (at 80 °C) | 23.5 | [58] |
| | N + PDDA–PAA | N/A | 0.086 | NA | [17] |
| | PAH–PAA | N/A | 0.068 | NA | [17] |
| | PPy/N/PPy | 3 | NA | NA | [59] |

PAA: Polyacrylic acid
PAH: Polyallylamine hydrochloride
PDDA: Polydiallyl dimethyl ammonium chloride PSS: Polyphenylsulfone
SPAEK: Sulfonated poly aryl ether ketone
Reproduced with permission from C.M. Branco, S. Sharma, M.M. De Camargo Forte and R. Steinberger-Wilckens, *Journal of Power Sources*, 2016, **316**, 139. ©2016, Elsevier [15]

filtration seem to offer better tolerance to methanol crossover as compared with composite membranes. Lin and Lu prepared a GO-laminated Nafion® membrane for DMFC application whereby the GO film prepared by filtration under gravity was transfer printed onto Nafion® 115 and then hot pressed together at 120 °C [61].

In a recent study, Bayer and co-workers prepared nanocellulose fibre and cellulose nanocrystal (CNC) membranes using vacuum filtration (Figure 3.3) for high-temperature fuel cell operation (100–120 °C). Dispersions of 3 and 11.8 wt% prepared in 500 mL of deionised water were vacuum filtered onto millipore filters (hydrophilic PTFE; pore size 0.1 μm). The resulting cellulose nanofibre (CNF) membranes were placed between two Teflon™ sheets before hot pressing for 20 min (at 110 °C and 1.1 MPa). The CNC membranes after filtration were simply pealed off the filter. The resulting free-standing CNF and CNC membranes had a thickness of 32 and 30 μm, respectively. The CNF were found to be flexible, but the CNC membranes were reported to be more brittle and breakable [66].

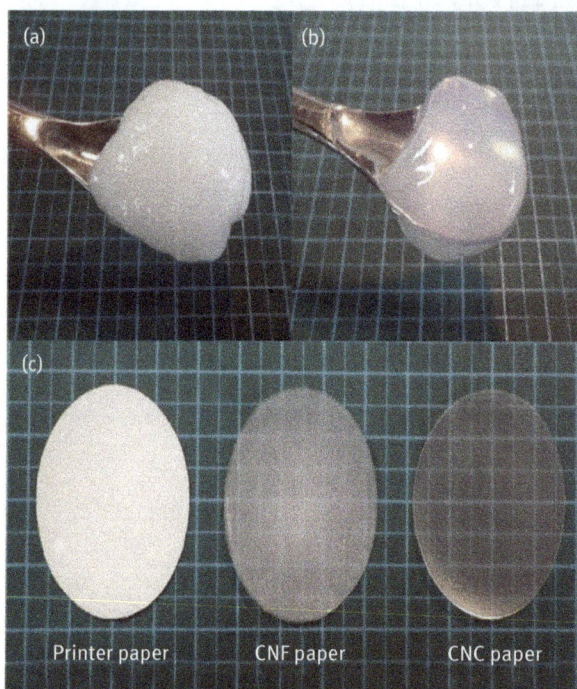

**Figure 3.3:** Photographs of a (a) CNF slurry, (b) CNC slurry, and (c) (left to right) conventional cellulose-based paper, CNF paper, and CNC paper (all similar thicknesses of ≈90 μm). Reproduced with permission from T. Bayer, B.V. Cunning, R. Selyanchyn, M. Nishihara, S. Fujikawa, K. Sasaki and S.M. Lyth, *Chemistry of Materials*, 2016, **28**, 13, 4805. ©2016, American Chemical Society [66].

## 3.5 Other approaches and concepts

Spin coating and spray coating are gaining interest for exploring the concept of multilayer membranes [67–70]. Lue and co-workers reported the drop coating and spin coating of dilute GO solution prepared in Nafion®/iso-propyl alcohol for formic acid, methanol and ethanol-based fuel cell systems [71]. Spray coating is also used commercially to produce barrier coatings for polymeric proton-conducting membranes. Spray coating has been explored to redesign the MEA concept whereby a GDL with electrocatalysts sprayed on it is spray coated further with thin layers of Nafion®. This process can reduce the thickness of membranes to nearly 10 μm. In another recent study, a multilayer-MEA assembly was prepared with gas diffusion electrode (GDE) spray coated with 200 nm of GO/cerium(IV) oxide (CeO$_2$) followed by spray coating of 10 μm of commercial Aquivion ionomer for PEMFC. The GO/CeO$_2$ interlayer reportedly reduced the hydrogen gas crossover current density to 1 mA/cm$^2$ [72].

Inkjet, screen and three-dimensional printing technologies are also being studied to improvise and revamp membrane and MEA concepts for fuel cells. The research team lead by Thiele has recently demonstrated interesting results using direct membrane deposition (DMD) utilising inkjet printing of a Nafion® layer (8–25 μm) directly onto anode and cathode GDE. They have also reported deposition of electrospun polyvinylidine fluoride-co-hexafluoropropylene directly onto GDE [73–75]. Direct deposition offers advantages such as reduced membrane thickness and ionic resistance without compromising mechanical strength because the membrane is supported on GDL. The authors also demonstrated that such a layer could mould perfectly because the anode/cathode surface morphology improved the adhesion between layers (Figure 3.4). The authors also suggested that such a printed membrane could also infiltrate and block the pores in the GDE, thereby

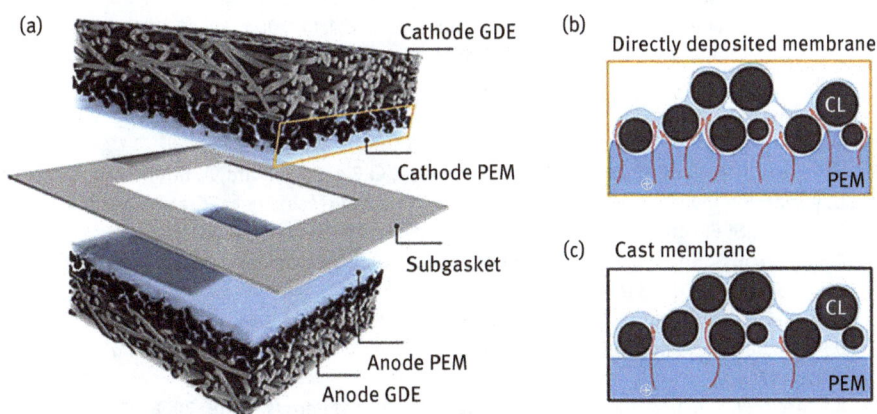

**Figure 3.4:** MEA preparation using DMD (schematic). (a) A thin polymer electrolyte membrane (PEM) layer is inkjet printed directly on anode and cathode GDE. A thin sub-gasket prevents hydrogen and current crossover through the end faces of the active area. (b) The dispersed polymer electrolyte (dark grey) can easily adjust to the CL surface, which leads to (c) relatively thin membranes and an increased electrolyte contact area of membrane, and the ionomer phase of the CL (light grey). Reproduced with permission from M. Klingele, M. Breitwieser, R. Zengerle and S. Thiele, *Journal of Materials Chemistry A*, 2015, **3**, 21, 11239. ©2015, The Royal Society of Chemistry [73].

improving contact between the catalyst layer (CL) and membrane and blocking paths for gas permeation. Reduced membrane thickness can also help to achieve better catalyst efficiency due to minimised resistance in the system. Spray coating may also be very effective for deposition of uniform layers in composite membranes with fillers (metal oxide or carbon nanomaterial), thereby allowing better size control of filler agglomerates along with uniform dispersion.

## References

[1]    J.D. Ferry, *Chemical Reviews*, 1936, **18**, 3, 373.
[2]    M.A. Hickner, H. Ghassemi, Y.S. Kim, B.R. Einsla, J.E. Mcgrath, M.A. Hickner, H. Ghassemi, Y.S. Kim, B.R. Einsla and J.E. Mcgrath, *Chemical Reviews*, 2004, **104**, 10, 4587.
[3]    M. Nakao and M. Yoshitake in *Handbook of Fuel Cells: Fundamentals, Technology, Applications*, Eds., W. Vielstich, H.A. Gasteiger and A. Lamm, John Wiley and Sons, Chichester, UK, 2003, **3**, 412.
[4]    V. Rao, N. Kluy, W. Ju and U. Stimming in *Handbook of Membrane Separations: Chemical, Pharmaceutical, Food and Biotechnological Applications*, Eds., Anil K. Pabby,S.S.H. Rizvi and A.M. Sastre, CRC Press, Boca Raton, FL, USA, 2008, p.568.
[5]    J. Thomassin, J. Kollar, G. Caldarella, A. Germain, R. Jérôme and C. Detrembleur, *Journal of Membrane Science*, 2007, **303**, 1, 252.
[6]    M. Mulder, *Encyclopedia of Separation Science*, 2000, **1967**, 3331.
[7]    B.S. Lalia, V. Kochkodan, R. Hashaikeh and N. Hilal, *Desalination*, 2013, **326**, 77.
[8]    Y. Liu, Y. Su, C. Chang, D. Wang and J. Lai, *Journal of Materials Chemistry*, 2010, **20**, 21, 4409.
[9]    H. Thiam, W.R.W. Daud, S.K. Kamarudin, A.B. Mohamad, A.A.H. Kadhum, K.S. Loh and E. Majlan, *International Journal of Hydrogen Energy*, 2013, **38**, 22, 9474.
[10]   R.F. Silva, De Francesco M. and A. Pozio, *Electrochimica Acta*, 2004, **49**, 19, 3211.
[11]   H. Jung, K. Cho, Y.M. Lee, J. Park, J. Choi and Y. Sung, *Journal of Power Sources*, 2007, **163**, 2, 952.
[12]   S. Vengatesan, E. Cho, H. Kim and T. Lim, *Korean Journal of Chemical Engineering*, 2009, **26**, 3, 679.
[13]   V. Salles, L. Seveyrat, T. Fiorido, L. Hu, J. Galineau, C. Eid, B. Guiffard, A. Brioude and D. Guyomar in *Nanowires – Recent Advances*, Ed., X Peng, InTech, Rijeka, Croatia, 2012, p.295.
[14]   D. Hassan, S. El-safty, K.A. Khalil, M. Dewidar and G. Abu El-magd, *Materials*, 2016, **9**, 9, 759.
[15]   C.M. Branco, S. Sharma, M.M. De Camargo Forte and R. Steinberger-Wilckens, *Journal of Power Sources*, 2016, **316**, 139.
[16]   W. Li and A. Manthiram, *Journal of Power Sources*, 2010, **195**, 4, 962.
[17]   S.P. Jiang and H. Tang, *Colloids and Surfaces A: Physicochemical and Engineering Aspects*, 2012, **407**, 49.
[18]   A.A. Argun, J.N. Ashcraft and P.T. Hammond, *Advanced Materials*, 2008, **20**, 8, 1539.
[19]   T. Yang, S.X. Zhang, Y. Gao, F.C. Ji and T.W. Liu, *The Open Fuel Cells Journal*, 2008, **1**, 4.
[20]   Q. Luo, H. Zhang, J. Chen, D. You, C. Sun and Y. Zhang, *Journal of Membrane Science*, 2008, **325**, 2, 553.
[21]   M. Marrony, J. Roziere, D. Jones and A. Lindheimer, *Fuel Cells*, 2005, **5**, 3, 412.
[22]   J. Lee, J. Won, K. Yoon, Y.T. Hong and S. Lee, *International Journal of Hydrogen Energy*, 2012, **37**, 7, 6182.

[23]  M.M. Hasani-Sadrabadi, E. Dashtimoghadam, N. Mokarram, F.S. Majedi and K.I. Jacob, *Polymer*, 2012, **53**, 13, 2643.

[24]  R. Padmavathi, R. Karthikumar and D. Sangeetha, *Electrochimica Acta*, 2012, **71**, 283.

[25]  Z. Shao, X. Wang and I. Hsing, *Journal of Membrane Science*, 2002, **210**, 1, 147.

[26]  B. Dong, L. Gwee, Salas-De La Cruz David, K.I. Winey and Y.A. Elabd, *Nano Letters*, 2010, **10**, 9, 3785.

[27]  H. Junoh, J. Jaafar, M. Noorul, A. Mohd, A.F. Ismail, M. Hafiz, D. Othman, M.A. Rahman, N. Yusof, W. Norhayati, W. Salleh and H. Ilbeygi, *Journal of Nanomaterials*, 2014, **2015**, 1.

[28]  N. Tucker, J. Stanger, M. Staiger, H. Razzaq and K. Hofman, *Journal of Engineered Fibers and Fabrics*, 2012, **7**, 63.

[29]  T. Tamura and H. Kawakami, *Nano Letters*, 2010, **10**, 4, 1324.

[30]  S. Subianto, S. Cavaliere, D.J. Jones and J. Roziere, *Journal of Polymer Science, Part A: Polymer Chemistry*, 2013, **51**, 1, 118.

[31]  H. Lin and S. Wang, *Journal of Membrane Science*, 2014, **452**, 253.

[32]  C. Lee, S.M. Jo, J. Choi, K. Baek, Y.B. Truong, I.L. Kyratzis and Y. Shul, *Journal of Materials Science*, 2013, **48**, 10, 3665.

[33]  S. Mollá and V. Compañ, *Journal of Membrane Science*, 2011, **372**, 1, 191.

[34]  I. Shabani, M.M. Hasani-Sadrabadi, V. Haddadi-Asl and M. Soleimani, *Journal of Membrane Science*, 2011, **368**, 1, 233.

[35]  W. Shan, P. Bacchin, P. Aimar, M.L. Bruening and V.V. Tarabara, *Journal of Membrane Science*, 2010, **349**, 1, 268.

[36]  S.U. Hong, R. Malaisamy and M.L. Bruening, *Langmuir*, 2007, **23**, 4, 1716.

[37]  C. Sheng, S. Wijeratne, C. Cheng, G.L. Baker and M.L. Bruening, *Journal of Membrane Science*, 2014, **459**, 169.

[38]  L. Ouyang, R. Malaisamy and M.L. Bruening, *Journal of Membrane Science*, 2008, **310**, 1, 76.

[39]  B.W. Stanton, J.J. Harris, M.D. Miller and M.L. Bruening, *Langmuir*, 2003, **19**, 17, 7038.

[40]  R. Malaisamy and M.L. Bruening, *Langmuir*, 2005, **21**, 23, 10587.

[41]  S. Mulyati, R. Takagi, A. Fujii, Y. Ohmukai and H. Matsuyama, *Journal of Membrane Science*, 2013, **431**, 113.

[42]  S. Chen, A. Bocarsly and J. Benziger, *Journal of Power Sources*, 2005, **152**, 27.

[43]  A. Peng, A. Morin, P. Huguet, Y. Lanteri and S. Deabate, *Physical Chemistry Chemical Physics*, 2014, **16**, 42, 23492.

[44]  Q. Wu, T. Zhao, R. Chen and L. An, *Applied Energy*, 2013, **106**, 301.

[45]  H. Dislich and P. Hinz, *Journal of Non-Crystalline Solids*, 1982, **48**, 1, 11.

[46]  L. Wang, B. Yi, H. Zhang, Y. Liu, D. Xing, Z. Shao and Y. Cai, *Journal of Power Sources*, 2007, **164**, 1, 80.

[47]  C. Lin, W. Lien, Y. Wang, H. Shiu and C. Lee, *Journal of Power Sources*, 2012, **200**, 1.

[48]  S. Zhong, X. Cui, C. Sun, S. Dou and W. Liu, *Solid State Ionics*, 2012, **227**, 91.

[49]  T. Yuan, L. Pu, Q. Huang, H. Zhang, X. Li and H. Yang, *Electrochimica Acta*, 2014, **117**, 393.

[50]  W. Liu, S. Wang, M. Xiao, D. Han and Y. Meng, *Chemical Communications*, 2012, **48**, 28, 3415.

[51]  S. Li, S. Zhang, Q. Zhang and G. Qin, *Chemical Communications*, 2012, **48**, 100, 12201.

[52]  C. Zhao, H. Lin, Q. Zhang and H. Na, *International Journal of Hydrogen Energy*, 2010, **35**, 19, 10482.

[53]  S. Yılmaztürk, H. Deligöz, M. Yılmazo lu, H. Damyan, F. Öksüzömer, S.N. Koc, A. Durmu and M.A. Gürkaynak, *Journal of Membrane Science*, 2009, **343**, 1, 137.

[54]  M. Yang, S. Lu, J. Lu, S.P. Jiang and Y. Xiang, *Chemical Communications*, 2010, **46**, 9, 1434.

[55]  H. Son, M. Cho, J. Nam, S. Cho, C. Chung, H. Choi and Y. Lee, *Journal of Power Sources*, 2006, **163**, 1, 66.

[56]  H. Lin, C. Zhao, W. Ma, H. Li and H. Na, *International Journal of Hydrogen Energy*, 2009, **34**, 24, 9795.
[57]  H. Lin, C. Zhao, W. Ma, H. Li and H. Na, *Journal of Membrane Science*, 2009, **345**, 1, 242.
[58]  M. Smit, A. Ocampo, M. Espinosa-Medina and P. Sebastian, *Journal of Power Sources*, 2003, **124**, 1, 59.
[59]  B. Yang and A. Manthiram, *Electrochemistry Communications*, 2004, **6**, 3, 231.
[60]  Ravikumar R. and K. Scott, *Chemical Communications*, 2012, **48**, 45, 5584.
[61]  C.W. Lin and Y.S. Lu, *Journal of Power Sources*, 2013, **237**, 187.
[62]  T. Bayer, S.R. Bishop, M. Nishihara, K. Sasaki and S.M. Lyth, *Journal of Power Sources*, 2014, **272**, 239.
[63]  Z. Jiang, Y. Shi, Z. Jiang, X. Tian, L. Luo and W. Chen, *Journal of Materials Chemistry A*, 2014, **2**, 18, 6494.
[64]  K. Sears, L. Dumée, J. Schütz, M. She, C. Huynh, S. Hawkins M. Duke and S. Gray, *Materials*, 2010, **3**, 1, 127.
[65]  C.D. Vecitis, M.H. Schnoor, M.S. Rahaman, J.D. Schiffman and M. Elimelech, *Environmental Science & Technology*, 2011, **45**, 8, 3672.
[66]  T. Bayer, B.V. Cunning, R. Selyanchyn, M. Nishihara, S. Fujikawa, K. Sasaki and S.M. Lyth, *Chemistry of Materials*, 2016, **28**, 13, 4805.
[67]  X. Zhu, H. Zhang, Y. Liang, Y. Zhang and B. Yi, *Electrochemical and Solid-State Letters*, 2006, **9**, 2, A49.
[68]  H. Li, M. Ai, F. Jiang, H. Tu and Q. Yu, *Solid State Ionics*, 2011, **190**, 1, 25.
[69]  J.L. Lu, Q.H. Fang, S.L. Li and S.P. Jiang, *Journal of Membrane Science*, 2013, **427**, 101.
[70]  H. Tang, G. Zhang and S. Ji, *AIChE Journal*, 2013, **59**, 1, 250.
[71]  S.J. Lue, Y. Pai, C. Shih, M. Wu and S. Lai, *Journal of Membrane Science*, 2015, **493**, 212.
[72]  M. Breitwieser, T. Bayer, A. Büchler, R. Zengerle, S.M. Lyth and S. Thiele, *Journal of Power Sources*, 2017, **351**, 145.
[73]  M. Klingele, M. Breitwieser, R. Zengerle and S. Thiele, *Journal of Materials Chemistry A*, 2015, **3**, 21, 11239.
[74]  M. Breitwieser, M. Klingele, B. Britton, S. Holdcroft, R. Zengerle and S. Thiele, *Electrochemistry Communications*, 2015, **60**, 168.
[75]  M. Breitwieser, C. Klose, M. Klingele, A. Hartmann, J. Erben, H. Cho, J. Kerres, R. Zengerle and S. Thiele, *Journal of Power Sources*, 2017, **337**, 137.

# 4 Methods used for membrane characterisation

## 4.1 Introduction

Characterisation studies are an integral part of any operating engineered system. These allow not only the identification of drawbacks and flaws in a material or a system, but also assist in achieving system optimisation. In the case of fuel cell-specific membranes, the advent of new and unique membrane materials and methods of membrane preparation demands the utilisation of existing as well as novel and innovative techniques to thoroughly scrutinise these membranes. Characterisation methods can be classified broadly into two categories: *ex situ* and *in situ*. *Ex situ* studies can provide specific information about the physical and chemical structures of the material as well as the prepared membranes. *In situ* studies facilitate understanding of the membrane behaviour and its responses to various stimuli generated inside an operating fuel cell system.

This chapter details the most commonly used *ex situ* and *in situ* methods used in membrane studies and characterisation. The *ex situ* methods have been discussed with reference to recent literature apart from explaining their concepts and operational details. *In situ* methods have been covered briefly with respect to conceptual and operational details since these have been elaborated upon using specific literature reports in Chapter 5.

## 4.2 *Ex Situ* characterisation methods

*Ex situ* studies aid identification and selection *via* elimination when exploring the suitability of a new material or preparation method for fuel cell membranes. They also help in developing an understanding of membrane behaviour under specific conditions in a simpler setup with fewer variables as compared with an *in situ* environment. The most widely investigated *ex situ* measurements are water uptake (WU), ion-exchange capacity (IEC), surface area, gas and fuel permeability with respect to crossover, chemical stability, and mechanical strength. Other material-characterisation studies, including X-ray diffraction, electron microscopy, and spectroscopic (X-ray photoelectron, Raman, and infrared) studies, are also regularly and extensively utilised to gain understanding regarding the uniformity and distribution (of fillers and pores) in an as-prepared membrane, along with chemical composition as well as other chemical and physical characteristics. Polymer and other materials should be studied comprehensively before using them for the preparation

Surbhi Sharma and Carolina Musse Branco

https://doi.org/10.1515/9783110647327-004

of membranes. Certain aspects, such as reorganisation in polymeric structures or the distribution of fillers in as-prepared membranes, can be understood better with these methods.

## 4.2.1 Water uptake

As discussed in previous chapters, water is essential to transport protons through the perfluorosulfonic acid (PFSA) membrane because the transport mechanisms (hopping and diffusion) are dependent on water. Thus, methods that evaluate how much water the membrane can hold outside the fuel cell form the basic tests in membrane research. Unlike the initial thoughts that the term WU might give rise to (i.e., the more water a membrane can hold the better), there is a limit up to which the WU can have a positive effect on membrane properties. A balance between the physical structure of the membrane and proton conductivity (PC) is the main challenge to achieve 'ideal' WU. PC is promoted by water due to the sulfonic groups that engage with the water molecules. Conversely, if the number of sulfonic groups is extremely high, the membrane becomes highly hydrophilic and may lose mechanical resistance or even dissolve in water in an extreme scenario.

The WU test was developed for all these reasons. Essentially, the test involves calculation of the difference between the dry weight ($W_d$) and wet weight ($W_w$) of the membranes, as shown in Equation 4.1 below, whereby the final result is given as a percentage:

$$WU = \left(\frac{W_w - W_d}{W_d}\right) \times 100 \qquad (4.1)$$

Although there is no standard procedure for this test, the most traditional way consists of cutting the membrane into small samples and soaking them in water for ≥24 h. The duration of soaking time changes according to membrane characteristics but, for Nafion®-like membranes, 24 h is considered sufficient. After this time, the membrane must be pat-dried and the wet weight recorded. Once the wet weight has been recorded, the samples are dried in an oven (under vacuum to expedite the process). The drying time also varies from membrane to membrane. High temperatures may be used for drying the membranes faster but high temperatures can affect the chemical structure of the membrane. Thus, it is important to define the temperature limit for the membrane under investigation. For Nafion®-like membranes, 80 °C for 24 h is usually enough to dry. For a novel membrane with new materials, the method to guarantee that the membranes are dried is to measure the weight of the membrane regularly at short intervals after drying for a few hours until there is no significant difference in the weight. It is recommended that each sample should be measured at least thrice at intervals of 30 min to obtain the average dry weight. Similarly, to record the wet weight, the sample should be soaked in water again for 30 min, and the weight recorded until consistent values are achieved. Table 4.1 details some of typical WU values reported for Nafions®.

**Table 4.1:** Typical values of water uptake for PFSA membranes.

| Membrane | Water uptake | Reference |
|---|:---:|---:|
| Nafion® NR-212 | ~24% | [1] |
| Cast Nafion® | ~22–27% | [1, 2] |

Another route to measure the WU is to dry the membrane first and record the dry weight first. Thereafter, soak the membrane and record the wet weight. However, the first route is preferred by most researchers. Besides the order of the measurement, temperature is also an important parameter. A few degrees of difference can significantly influence the amount of water held in the membrane [3, 4]. Therefore, all measurements must be carried out at the same temperature. Conversely, considering that, in a real fuel cell, the membrane is not at room temperature (RT), the water in the *ex situ* WU test can be heated to the desired temperature while following the same WU test procedure. For obvious reasons, intermediate-temperature (IT) and high-temperature (HT) proton-exchange membrane fuel cell (PEMFC), i.e., IT-PEMFC and HT-PEMFC, environments cannot be exactly reproduced in this kind of test.

It is important to notice that the two routes discussed above represent the water that the membrane can hold in the presence of liquid water. Another method has been developed to evaluate the WU in the presence of water vapour. Here, in a chamber with a heating system, the water is vapourised and the membrane is submitted to it. A humidification controller is used to control the amount of water. This test is more analogous to the environment inside the fuel cell. It is, however, a time consuming and more complex method for a simple test because the information obtained from WU is only indicative of the membrane's fuel cell performance. As a counter argument, it is necessary to add to the complexity of the *ex situ* test to evaluate if the membrane is worthy of expensive, time-consuming *in situ* fuel cell testing, especially if investigating a novel membrane material.

Figure 4.1 shows the WU behaviour reported for membranes with graphene oxide (GO) and sulfonated graphene oxide (SGO) fillers as reported by Beydaghi and co-workers. They reported crosslinked-nanocomposite membranes prepared from polyvinyl alcohol (PVA) and aryl sulfonated GO using gluteraldehyde (GLA) as a crosslinking agent [5]. The researchers dried membranes at 60 °C before weighing them and, for wet weight, soaked the membrane in deionised water for 12 h at RT.

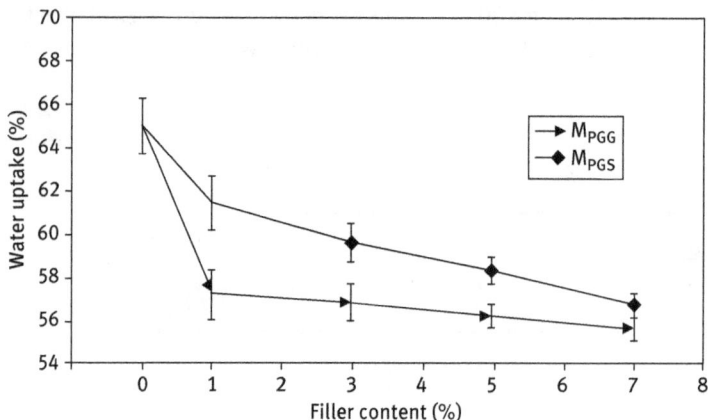

**Figure 4.1:** WU of nanocomposite membranes PVA/GLA/GO (MPGG) and PVA/GLA/SGO ($M_{PGS}$). Reproduced with permission from H. Beydaghi, M. Javanbakht and E. Kowsari, *Industrial & Engineering Chemistry Research*, 2014, **53**, 43, 16621. ©2014, American Chemical Society [5].

### 4.2.2 Ion-exchange capacity

IEC is defined as the capacity of a material (the membrane, in this case) to swap/exchange the ions that are attached to and loosely incorporated in the structure with the ions that are within the surrounding solution without the material itself dissolving.

The membranes usually consist of sulfonic, carboxylic acid or amino/ammonium groups. These groups are responsible for ion exchange. In the case of PFSA membranes, the active groups for ion exchange are the sulfonic groups. The ability of a membrane to exchange these ions helps in understanding its ability to transport protons inside a PEMFC. However, IEC is only an indicative test for fuel cell application. In general, membranes that show low IEC values also show low PC. Conversely, membranes with high IEC do not necessarily demonstrate high PC because the PC inside the fuel cell is much more complex and depends on a wide range of factors.

The method to define the dissociation of ions from the membrane and exchange with the ions in the solution is pH titration with a salt solution. IEC is calculated by determining the quantity of membrane counter ions necessary to reach a neutral pH upon ion exchange with a salt solution. The IEC is given by milligram equivalent per gramme of the membrane (mequiv/g) [6].

To carry out the IEC test, the membrane dry weight is measured before immersing it in the saturated salt solution for ≥72 h to allow ion exchange (Figure 4.2). Sodium chloride is usually employed. Thus, the hydrogen ions ($H^+$) in the membrane are exchanged with the sodium ions ($Na^+$) in the solution. The proton ($H^+$) release

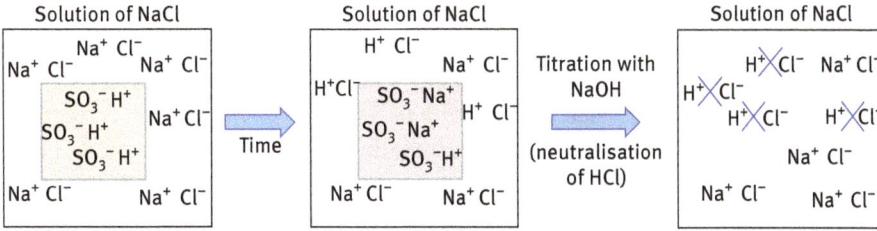

**Figure 4.2:** Schematic showing the ion exchange between the membrane and salt solution with further neutralisation with the alkaline solution.

is evaluated by titration and neutralisation using a standard alkaline solution of known concentration. The most common alkaline solution is sodium hydroxide. The IEC is determined using Equation 4.2, where $V_X$ and $M_X$ are the volume and concentration of the alkaline solution titrated X, respectively, and $W_d$ is the dry weight of the membrane [7–9]:

$$IEC = \frac{V_X \times M_X}{W_d} \tag{4.2}$$

Although this is by far the most common route, some scientists, such as Tohidian and co-workers, prefer to follow the inverse route and immerse the membrane in an alkaline solution and titrate with an acid solution [10].

IEC and WU influence each other. For PFSA membranes, the IEC increases with WU [11]. As explained above, the WU increases with temperature. Thus, it is recommended that all samples should be measured together under a controlled environment. Another important aspect is that titration needs an indicator to define when the solution is neutralised. For acid–base titration, the most commonly used indicator is phenolphthalein. Typically, IEC is carried out for the same sample, which is subjugated to the WU test, which means that value of dry weight has been already measured.

As discussed in Chapter 2, another important property for sulfonated membranes is the hydration number ($\lambda$), which is defined as the number of water molecules per sulfonic group. The '$\lambda$' can be calculated by Equation 4.3, where $M_{H2O}$ is the water molar mass, and '$W_w$' and '$W_d$' are the wet and dry weights of the membrane, respectively:

$$\lambda = \frac{W_w - W_d}{IEC \times W_d \times M_{H_2O}} \tag{4.3}$$

The hydration number is an important property because it considers IEC and WU. It can be considered to be the 'efficiency' of sulfonic groups available in the membrane. The membrane can have low sulfonation, but if each sulfonic group attracts a high number of water molecules, the final membrane will be capable of high proton transport. On the other hand, if each sulfonic group in a membrane attracts a small number of water molecules, the polymer must be highly sulfonated to achieve average PC behaviour. The high degree of sulfonation (DS) in such a membrane eventually results in mechanical stability issues.

Besides this, the hydration number is important to determine the speed of proton transport, which is faster with high values of hydration numbers [12]. The value of the hydration number defines which transport mechanism is preferred as well as the increase or decrease in energy activation. For example, the Grotthuss mechanism is dominant if $6 \leq \lambda \leq 15$ at low temperatures [13].

### 4.2.3 Gas permeability

Gas permeability studies are essential to understand membranes prepared using new materials in composite as well as multilayered structures. An 'ideal' electrolyte should be 'gas tight' or gas impermeable. Most membrane materials however, are not completely impermeable to gases and allow a certain degree of crossover of oxygen as well as hydrogen across to the other side. Certain composite and multilayer structures (using graphene, silica or other filler materials) show reduced gas permeation. Typically, a gas-permeability test measures the volume of gas passing through the sample when driven by a certain pressure gradient in a given time (Equation 4.4):

$$K = \frac{I}{\Delta_p} - \frac{\Delta V}{\Delta \tau} \left[ \frac{2p_2}{(p_1 + p_2) \, \Delta p} \right] \tag{4.4}$$

where 'K' is the permeability, '$\Delta p$' is the pressure drop (difference between '$p_1$' and '$p_2$'), $\Delta V$ is the change in volume, '$\Delta t$' is the time period for which the permeability is studied, and 'I' is a constant.

In a recent study on cellulose nanofibres (CNF) and cellulose nanocrystals (CNC), Bayer and co-workers performed gas permeation studies by masking the membrane area with Kapton® and alumina tape to provide a circle of required diameter (d = 1 cm) and area (S = 0.785 cm$^2$) and additionally, placed on a porous polycarbonate support filter (pore size 1.2 μm) to prevent membrane deformation during measurements. The dry hydrogen permeation rate through the two cellulose membranes and a Nafion® 112 membrane was measured between RT to 80 °C using a GTR-11A/31A gas barrier testing system (GTR Tec, Japan) utilising a differential pressure method. The system induces gas permeation by vacuum on the permeate side and extra pressure applied at the feed side [14]. Further specifications of the test include the following total pressure difference of 200 kPa, and 30 min of sample collection time after vacuuming the sweep side of the membrane. The collected gas volume is measured after it has transferred to a gas chromatograph fitted with a thermal conductivity detector.

Gas permeation tests for nanocellulose-based membranes revealed hydrogen permeability that was nearly three orders of magnitude lower than that of Nafion® (Figure 4.3). The permeability however, increased with temperature. The exceptionally low permeability of nanocellulose membranes was associated with a significant decrease in the free volume compared with conventional cellulose materials (with micron-scale fibre diameters) due to the dense packing of the individual nanoscale fibres/crystals. The low permeability values also suggested the membranes to be highly uniform and with minimal structural defects [15].

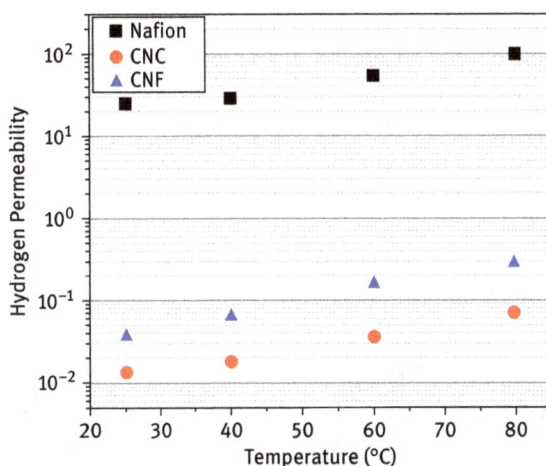

**Figure 4.3:** Hydrogen permeability of CNF and CNC membranes in comparison with Nafion® NR-212. Reproduced with permission from T. Bayer, B.V. Cunning, R. Selyanchyn, M. Nishihara, S. Fujikawa, K. Sasaki and S.M. Lyth, *Chemistry of Materials*, 2016, **28**, 13, 4805. ©2016, American ChemicalSociety [15].

## 4.2.4 Methanol permeation test

Methanol mixes instantly with water and, thus, easily penetrates into the proton-conduction membrane (PCM). Inside the membrane, methanol can combine and react with the migrated electrocatalysts to accelerate the degradation process. In another scenario, methanol can pass through the membrane entering the cathode and interfere with cathode reaction in addition to electrocatalyst poisoning. Additionally, methanol passing through the membrane is not being used as fuel and is resulting in loss of energy capacity. Consequently, methanol crossover measurements are very important for evaluation of direct methanol fuel cell (DMFC) performance.

The common methods adopted to minimise or prevent methanol crossover completely are to increase the tortuosity of the pathway inside the membrane, increase the membrane thickness or improve membrane selectivity. The electrochemical selectivity of a membrane will depend on the polymer/material and is defined as the ratio between PC and methanol permeability. The former two approaches are known to negatively affect the PC of the membrane as well. Consequently, the best approach to achieve reduction in methanol crossover without interfering with its PC behaviour is through enhancing it selectivity [16].

Measurement of the crossover or permeability measurement is carried out through an *ex situ* test. A diaphragm diffusion cell (usually made of glass) consisting of two compartments is used. The membrane to be tested is placed between two equal compartments. One of the compartments is filled with methanol solution of known concentration while water is filled in the other (Figure 4.4). Methanol flux generated across the membrane due to the concentration gradient between the two sides results in crossover and, after specific time intervals, the concentration of methanol is measured in both compartments to detect the crossover. This test can be carried out either at RT or at elevated temperatures.

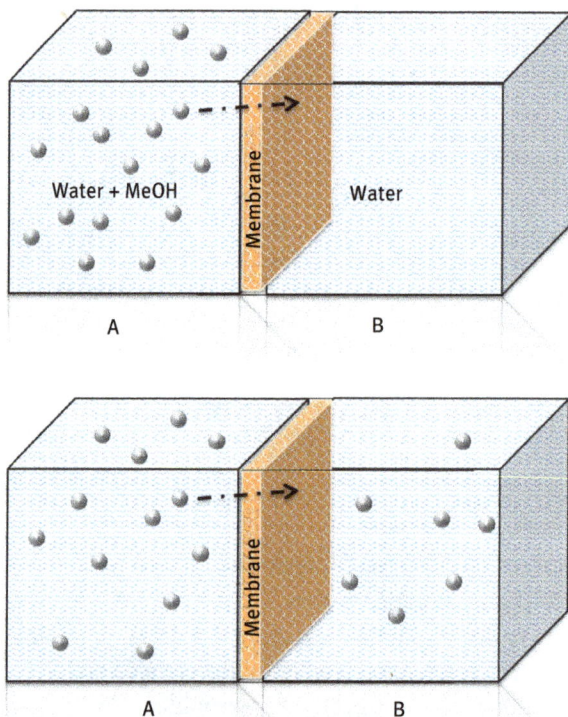

**Figure 4.4:** Methanol permeability test cell at time zero (top) and after time interval X (bottom).

Tripathi and Shahi reported methanol permeability studies of the composite membranes *N-p*-carboxy benzyl chitosan–silica–PVA hybrids. They used a two-compartment diaphragm diffusion cell with each compartment having a volume of 50 cm³. The compartments were separated using a vertical membrane with an effective area of 20 cm². The membrane was clamped between the two compartments. Prior to testing, the membranes were equilibrated in a water–methanol mixture for 12 h. Initially, one compartment was filled with a 30 or 50% (v/v) methanol–water mixture while another with double-distilled water. Both compartments were stirred continuously. Measuring the refractive index using a digital refractometer enabled the monitoring of the increase in methanol concentration with time in the compartment initially filled with double-distilled water [17]. The methanol permeability 'P' finally was obtained using Equation 4.5:

$$P = \frac{1}{A} \frac{C_{MEOH-A}(t)}{C_{MEOH-A}(t - t_0)} V_B L \tag{4.5}$$

where 'A' is the effective membrane area, 'L' is the thickness of the membrane, '$C_{MeOH-B}(t)$' is the methanol concentration in compartment B at time 't', $C_{MeOH-A}$ $(t - t_0)$ is the change in methanol concentration in compartment A between time '0' and t, and '$V_B$' is the volume of compartment B [17].

Methanol permeability can also be determined in DMFC as an *in situ* test. The standard membrane electrode assembly (MEA) preparation and fuel cell setup should be made, except on the cathode, which must be fed with nitrogen. The crossover value is obtained by recording the limiting oxidation current of methanol at the cathode with open circuit voltage (OCV). Essentially, the amount of methanol crossover from the anode to the cathode is determined on the basis of the mass balance between the amount of methanol supplied to the cell, the amount of methanol consumed during the electrochemical reaction, and the amount of methanol collected in the outlet. The desired concentration of methanol (usually 2 M) is fed to the anode while maintaining the operating temperature (OT) of the fuel cell. As the cell runs for a few hours, unreacted methanol is collected from the anode outlet. Samples of methanol supplied at the inlet and collected at the outlet can then be measured for their density to determine the methanol concentration. Permeability can be determined by calculating the difference between the amount of methanol supplied at the inlet and the amount of methanol collected at the outlet [18, 19]. This method represents a more realistic assessment of the membrane performance towards methanol permeability as compared with *ex situ* studies. However, the *ex situ* crossover test is more commonly used due to the ease of setup and the test itself being quicker.

### 4.2.5 Chemical degradation and durability testing

PFSA and other polymer-based membranes can be tested using specific *ex situ* methods to study their chemical durability and degradation mechanism. These are based on Fenton's reactions where hydrogen peroxide ($H_2O_2$) is reduced to the hydroxyl ($OH^\bullet$) radical and hydroperoxyl ($OOH^\bullet$) radical in a catalytic cyclic oxidation–reduction reaction with cationic ($Fe^{2+}$ and $Fe^{3+}$) iron as the catalyst, as shown in Equations 4.6 and 4.7 [20–22]:

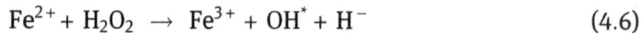

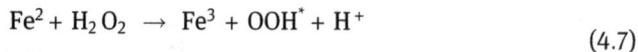

$$Fe^{2+} + H_2O_2 \rightarrow Fe^{3+} + OH^* + H^- \tag{4.6}$$

$$Fe^2 + H_2O_2 \rightarrow Fe^3 + OOH^* + H^+ \tag{4.7}$$

These Fenton reaction-based tests are essentially contaminant degradation studies for polymer membranes. These are designed to mimic the conditions created inside the operating fuel cell due to chemical and electrochemical reactions taking place at the electrocatalytic surface with crossover gases and the dissolution of iron and other multivalent ions due to corrosion and degradation of other cell materials (such as metallic bipolar plates, electrocatalysts) over a long time [23]. Such tests are, therefore, referred to as accelerated chemical degradation tests. The reactive radical species produced in the above reactions react with the side chains of PFSA and other polymers with hydrogen-containing terminal bonds. Other metal ions, including $Cu^{2+}$, $Ni^{2+}$, $Co^{2+}$, $Cr^{3+}$, $Na^+$, $K^+$, $Ca^+$ and $Li^+$, are also used as catalysts for Fenton's reactions to generate the radical species because these metals are commonly used in electrocatalysts or end plates/bipolar plates of PEMFC systems. These tests are also a useful tool for understanding membrane degradation mechanisms, which are yet to be completely comprehended. These tests can be done as *ex situ* or *in situ* studies. Of course, the conditions inside a fuel cell cannot be replicated completely in an *ex situ* setup. Parameters such as relative humidity (RH), OT, gas type, flow rate and pressure and cell voltage have significant roles in defining *in situ* studies.

In general, two approaches are often reported for carrying out such durability studies. The first approach uses a solution of $H_2O_2$ ($\approx$30 wt%) containing a trace amount of $Fe^{2+}$ (4–20 ppm) [24, 25]. The second approach uses an ion-exchange process in which acid sites of the membrane are replaced with metal ions before the membrane is exposed to $H_2O_2$ [26]. Some modified versions of the test include the test solution being bubbled with $H_2$ and air in different ratios to simulate anode/cathode environments or the use of electric fields applied to the test solution [27, 28]. In any case, the residual solutions (or excess water collected from the cathode in case of *in situ* tests) can be extensively evaluated for rates of fluoride ion release (also referred to as the fluoride emission rate using high-performance liquid chromatography and an ion-selective electrode). The degraded, post-test (or in the case of an *in situ* end-of-life) membrane or MEA can also be investigated for changes in chemical composition and structure using liquid or solid-state nuclear magnetic resonance (NMR), mass spectroscopy (MS), Fourier-transform infrared (FTIR) spectroscopy and ion chromatography. Electron microscopy and X-ray photoemission spectrometry (XPS) can also be very useful for studying the chemical, mechanical and structural changes in membranes.

Several studies have explored the durability of Nafion® membranes using these tests at different temperature and RH conditions for simulated anode as well as cathode environments for variable times (few hours to days) [24–26]. Nafion® and Nafion®-composite membranes subjected to Fenton's test conditions reveal significant polymer chain degradation *via* loss of C–F and sulfonic acid bonds. Studies by Kinomoto and co-workers revealed a loss of 68% and 33% for C–F and sulfonic acid bonds over an exposure (immersion) period of 9 days, respectively [26]. FTIR spectroscopy studies reported by Tang and co-workers observed a loss of stretchability in CF2, S–O and C–O–C bonds with increasing exposure time, and concluded that Nafion® degradation initiates from the ends of the main polymer chain, leading to a loss of chain-repeating units and pinhole formation [29].

More recently, membranes with other polymers have also been put to test using Fenton's tests [30–32]. Wong and Kjeang carried out Fenton's test on polybenzimidazole (PBI) membranes [33]. They observed that Fe ions within the MEA resulted in a redox cycle sustaining to a higher concentration of Fe ions and resulting in severe degradation at OCV. In a further study, Chang and co- workers developed a new *ex situ* method and carried out an 'electro- Fenton test' to study the effect of an electric field on the oxidative degradation of PBI [28]. Their setup consisted of a crystallising dish with 30 wt% $H_2O_2$ and 20 ppm of Fe(II) ions in which the PBI membrane was immersed. The dish was placed in a preheated oven at 68 °C with the Fenton solution replaced every 16 h. The dish was connected to a direct current (DC) voltage supply with copper plates, and electric-field intensities of 5,000 and 2,000 V/m were set using the DC voltage, separation between the copper plates and the height of the Fenton reagent. The electric-field intensities were evaluated to be at 0.95 V (5,000 V/m) and 0.0.60 V (2,000 V/m) based on COMSOL calculations. They characterised the degraded membranes extensively using scanning electron microscopy, FTIR spectroscopy, $^1$H-NMR and XPS. The degraded membranes were found to show loss of weight, intrinsic viscosity and thermal stability as the electric field increased. Using XPS they also identified that the electric field accelerated the formation of amide and amino groups after breakdown of the benzimidazole rings in PBI.

Salleh and co-workers prepared sulfonated polyether ether ketone (SPEEK)/ Cloisite®/triaminopyrimidine (TAP) nanocomposite membranes and studied their durability for use in DMFC operation in comparison with Nafion® [31]. They used four Fe ion concentrations (0.8–50 ppm) and four time durations (12–96 h). Combining density functional theory with FTIR characterisation, they revealed that the C–O–C and –$SO_3H$ bonds with a phenylene ring, and hydrogen bonds between SPEEK, Cloisite®, and TAP were prone to radical attack. Loss of these functional groups further resulted in structural deformation, loss of mechanical strength, and reduced hydrophilicity in the nanocomposite membrane. The composite also fared poorly against Nafion® due to loss in WU and PC. However, the membrane selectivity of the SPEEK/Cloisite®/TAP nanocomposite membrane was found to be higher (27,037 S•s/cm$^3$) than that of Nafion® (3,292 S•s/cm$^3$) because it could retain higher methanol barrier properties after radical-attack degradation. Interestingly, they also developed a correlation between Fenton's test and the DMFC lifespan of the membrane (which is usually difficult to evaluate because the $H_2O_2$ and Fe ion in the MEA will depend on multiple factors in an operating system). They reported their membrane's estimated lifespan to be 9,800 h.

Liu and co-workers reported a study of the degradation of sulfonated poly aryl ether ketone (SPAEK)-containing alkyl sulfonated side chains using Fenton's test. Their systematic study for 10 exposure times from 0–45 h was characterised extensively using $^1$H-NMR, FTIR spectroscopy, gel permeation chromatography, thermogravimetric analysis and atomic force microscopy along with PC and elemental analysis. Their results suggested that sulfonic acid groups remained relatively stable during Fenton's test. However, the main chains of the polymer were disrupted significantly and the apparent structural degradation of the SPAEK was ascribed mainly to cleavage of the main polymer chains [32].

Apart from Fenton's test, ultraviolet (UV)-assisted photocatalytic breakdown of $H_2O_2$ is utilised to study membrane degradation. This method exposes the membrane to the reactive hydroxyl and hydroperoxyl radicals in the absence of metal ions. Alsheheri and co-workers prepared aromatic trifluoromethyl sulfonamide compounds (mono-, di-, and tri-perfluorinated sulfonamide) as model compounds for the study of perfluorosulfonamide polymers and studied their stability in Fenton's reagent and UV-irradiated $H_2O_2$ [30]. In the UV-assisted degradation process, 2 moles of radicals are produced for each mole of photolysed $H_2O_2$. In their setup, Alsheheri and co-workers purged 10 mM solution of the model compounds with dry $N_2$ gas for 10 min before mixing it with an equivalent concentration of $H_2O_2$. The solution was then placed ≈15 cm away from a UV light source for 1 h at RT with constant stirring. They used a 170 W mercury UV lamp (intensity ≈500 ± 40 mW cm$^{-2}$) with a broad-irradiation spectrum wavelength range of 200–2500 nm. Based on their post-degradation analysis after Fenton's test using liquid chromatography/MS equipped with electrospray ionisation detection in negative mode, they suggested two pathways of degradation for the model compounds. According to the first pathway, ring fusion and chain oxidation takes place in the case of the mono-substituted compounds as interpreted from the MS peaks at 499 and 249 Da, which can be assigned to adduct formation as a result of fusion of a trifluoromethyl sulfonamide chain to the aromatic ring. The peak at 249 kD was attributed to doubly ionised species. The second pathway suggested the conversion of the model compound chain end group, -CF$_3$, into -COOH. The mechanism proposed for this degradation is similar to that of the attack of radicals on the polytetrafluoroethylene (PTFE) backbone of Nafion® and other commercial polymer electrolyte membrane (PEM) polymers, which begins with a radical attack at the -CF$_3$ end chain group to form the radical species R-CF$_2$, which in turn captures a second hydroxyl radical to produce the R-CF$_2$-OH intermediate species. On the other hand, their analysis of UV-assisted degradation resulted in a species which was attributed to a combination of fusion of two rings of the mono model compounds, oxidation of one sulfonamide chain, and multiple ring hydroxylations. They based this hypothesis upon the peaks observed at 599, 615, and 630 Da, which the authors attributed to tri-, tetra-, and penta-hydroxylated products. UV-assisted degradation also resulted in a species with MS peaks at 304 and 320 Da, which the authors accredited to the direct hydroxylation of the mono-substituted compound. They concluded that perfluorinated sulfonamides tend to exhibit chemical degradation similar to that of the perfluorinated ether polymers currently in use in PEMFC [30].

## 4.2.6 Tensile test

Mechanical stability of the membranes is usually studied using a standardised tensile test, which is used for plastics and resin-like materials. The standard method, also referred to as American Society for Testing and Materials (ASTM) D638, uses a universal tensile-compression machine and requires the test specimen to be dumb-bell-shaped with a gauge length of 50 mm (or 25 mm). To carry out the test, membrane test specimens are cut into (70 × 25 mm) strips. The ideal thickness of the test specimen must be 3.2 ± 0.4 mm (or else ≤7 mm). Specific test conditions of RH, temperature, and crosshead speed can have a significant effect on the test results. The test specimen is placed between the grips of the machine and subjected to elongation at the set speed and force. The tensile strength (TS) can be calculated using the load (at which the membrane breaks) and the membrane cross-section area as shown in Equation 4.8:

$$\sigma = \frac{L}{width_G \times thickness} \tag{4.8}$$

where 'σ' is the TS in MPa, 'L' is the load (in Newtons) at which the membrane breaks, '$width_G$' is the gauge width in mm, and 'thickness' is the membrane thickness in mm.

Information on the elastic modulus, yield stress, yield strain, stress and strain at break can be obtained using the load–elongation curves obtained experimentally. These data can also be converted further into stress–strain plots [31, 34]. In an interesting study, Reyna-Valencia and co-workers studied the mechanical properties of SPEEK membranes along with a) the effect of DS and b) presence of the filler boron orthophosphate (BPO4) in a SPEEK membrane (where the studied $BPO_4$:SPEEK ratios were 15:85, 30:70 and 50:50). They studied the membranes at different temperatures (from 23 to 40 °C) and at 30% RH as well as after soaking them in water. Their tensile test arrangements included load cells of 100 N and 1 kN, crosshead speed of 5 mm/min and gauge width of ≈10 mm, while the membranes tested were 70–200-μm thick. Their studies concluded that 69% sulfonated SPEEK (in comparison with 63 and 83% sulfonated SPEEK) demonstrated the best mechanical properties.

The tensile properties for some of the membranes reported in the literature are shown in Table 4.2.

**Table 4.2:** Compilation of reported TS for different membranes reported in literature.

| Membrane | Test conditions | Tensile properties | | Ref. |
|---|---|---|---|---|
| | | Young's modulus (MPa) | Yield strength (MPa) | |
| Nafion® 117 | 25 °C; RH: 30<br>25 °C; RH: 90<br>85 °C; RH: 30<br>85 °C; RH: 90<br>Strain rate: 0.2 mm/mm per min; along transverse direction | 197<br>121<br>121<br>46 | – | [35,36] |
| Nafion® 212 | 25 °C; fully hydrated membrane; crosshead speed at 50 mm/min | 112.6 | 9.0 | [37] |
| Sulfonated polyarylene ether sulfone/ polyacrylonitrile (electrospun non-woven) DS: 50 | 25 °C; fully hydrated membrane; crosshead speed at 50 mm/min | 692.1 | 13.8 | [37] |
| Aquivion E87-05S E98-05S | ASTDM D882; 23 °C 50% RH | – | 30 | [38] |
| Gore® Select–expanded-PTFE reinforced | 25 °C; 50% RH | 560 | 19 | [39] |
| PBI/radiation-grafted GO/ phosphoric acid | 10 mm/min for TS; 2 mm/min for tensile modulus | – | 36.4 | [40] |
| Nafion/CeO$_2$ | 23 °C; 44% RH 5 mm/min strain rate | 203 (CeO$_2$ 3 wt%)<br>207 (CeO$_2$ 5 wt%)<br>209 (CeO$_2$ 10 wt%) | – | [41] |
| SPEEK (DS: 60–65%) | 1 mm/min | – | 47 | [31] |
| SPEEK/Cloisite®/TAP | 1 mm/min | – | 51 | [31] |

Another commonly used test for examining mechanical durability is RH cycling. *Ex situ* RH cycling involves the use of an environmental chamber in which the membrane can be exposed to humidified gases with alternate low and high RH, which gives rise to mechanical stress in the membrane due to cyclic shrinking and expansion/swelling. Temperature can also be set and fracture toughness is measured under variable temperature and RH conditions. Selected studies have explored PFSA membranes using RH cycling and revealed interesting trends, such as increase in gas crossover with increase in RH amplitude, which was attributed to large swelling/shrinkage stresses; reduced crack propagation and delayed crack/craze formation in expanded PTFE-reinforced membranes [29, 42, 43]. Patankar and co-workers reported using a knife slit test combined with RH cycling and variable temperatures on three commercial PEM (Nafion® 211, Nafion® 111-IP and Gore® Select 57). At each temperature/RH test condition, the membranes were tested with cutting rates of 0.1, 1, 10, 50 and 100 mm/min. Thus, the fracture energy associated with the knife slit process was doubly shifted using temperature and humidity shift factors to construct the master curves. Based on the shift with respect to temperature and humidity, the authors suggested that the slitting process is viscoelastic in nature. The effect of humidity on fracture energy was found to be less pronounced at higher temperatures, which is most relevant to operation of automotive fuel cells. They also observed that although the pinhole formation propagated at high temperatures, it is accelerated even more by increasing the temperature instead of RH. Their results indicated that, at a given fracture energy, the corresponding reduced cutting rate for Nafion® NR-211 could be 10–100-fold faster than those associated with Nafion® 111-IP or Gore® Select 57 membranes [44].

### 4.2.7 Thermal stability and thermomechanical durability tests

Thermal stability of membranes is studied with the help of freeze/thaw (F/T) cycles and sub-zero start-up. These are mostly done using *in situ* tests. If carried out as *ex situ* studies, these can be carried out on the membrane or MEA. The nature and properties of the PCM imply that thermal stability cannot be measured independent of the response with reference to water content or mechanical properties. As such, thermal stability studies are often considered part of mechanical stability studies. Indeed, many studies have explored the thermomechanical behaviour of PCM. The *ex situ* test for a membrane alone or using MEA can be carried out using specialised temperature cycling chambers usually capable of exposing the membrane/MEA to hundreds of F/T cycles (usually between −40 to +80 °C). Changes to mechanical and physical characteristics such as TS and equivalent weight are studied before and after exposure. Changes to membrane oxygen permeability and gas crossover can also be measured using permeability apparatus and a glass eudiometer, respectively. Thermomechanical tests can also be done in dynamic mechanical analysis chambers, which not only cycle temperature but also loading paths. For MEA, polarisation plots and hydrogen adsorption/desorption tests can also be carried out [45, 46]. Chen and co-workers reported *ex situ* thermomechanical studies on

Nafion® NR-212 (thickness: 0.0503 mm) using an ASTM standard with 25 × 5 mm rectangular strips. They studied the responses of membranes to three thermomechanical-coupling loading paths: proportional path (i.e., thermal and mechanical load applied in phase), rectangular path in a counter-clockwise direction, and rectangular path in a clockwise direction. The temperature range was −30 to +55 °C (rate of temperature change: ±8 °C) and the stress ranged from 0 to 8 MPa (rate of stress change: ±2 MPa). Their work revealed that maximum creep did not occur at the lowest temperature. A history dependence was observed for thermal stresses, and the effect of initial temperature on the creep in the initial cycle played a substantial part as evidenced by significantly higher creep strains for the counter-clockwise direction loading path as the temperature decreased from +55 to −30 °C as opposed to the clockwise direction path (from −30 to +55 °C). Their study also identified that the strain on a specimen was composed of two parts: the accumulated plastic strain due to cyclic stresses and the creep strain in each cycle [46].

RH and F/T cycling is discussed in further detail with reference to *in situ* tests reported in recent literature in Chapter 5.

## 4.3 *In situ* characterisation

Single-cell systems are commonly used for *in situ* studies in fuel cells. This is the simplest test system that can be studied on a laboratory scale, and can provide performance information specific to any operating fuel cell stack in a specific environment and application. Single-cell systems and 5–10 cell stacks are also operated for long duration (hours and days) under higher stress conditions (referred to as 'accelerated stress tests') to simulate and evaluate the performance of fuel cell components over the long-term and if used for a specific application.

### 4.3.1 Proton conductivity

Among the tests realised before the real fuel cell-performance test, PC is probably the most important. It shows the ability of the membrane to transport protons through- or in-plane. PC values can also be achieved from an *ex situ* impedance measurement. However, the advantage of this test when compared with an impedance test is that the exact fuel cell conditions can be reproduced in the test. If a membrane does not show high or at least moderate PC, it is almost impossible to show an acceptable fuel cell performance. However, if the PC is high it does not necessarily mean that the performance in the PEMFC will be excellent because of the many parameters inside the fuel cell, as discussed in previous sections. In DMFC, the performance is more independent of PC. Methanol crossover is an important factor which strongly influences the overall performance. Methanol, unlike $H_2$, mixes instantly with water. Water is necessary for proton transport inside the membrane, so methanol traverses the membrane with water and reaches the cathode to affect electrocatalyst behaviour. Thus, for DMFC, selectivity is the most important property (i.e., the relationship between methanol permeability and PC) [47].

There are two main parameters and variances to decide before proceeding with the PC test: the direction of the measurement and the number of probes. The direction of measurement denotes whether the PC will be analysed through- or in-plane with respect to the membrane (Figure 4.5). The advantage of the through-plane measurement is that the entire membrane cross-section is relevant. This method is more similar to the pathway that the protons must go through when the membrane is fitted in an operating fuel cell. Therefore, the results for PC behaviour are more similar to those inside the fuel cell. On the other hand, in-plane measurements analyse the conductivity in the surface direction, which does not necessarily represent the direction in which the protons are transported. Both tests are relevant if investigating anisotropic membrane materials.

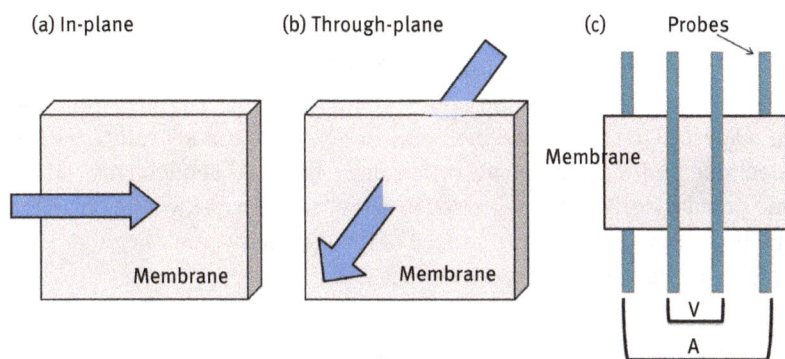

**Figure 4.5:** Possible directions (arrows) of PC measurement, a) in-plane and b) in the right through-plane. c) PC design for the four-probe method, with setup in-plane.

### 4.3.1.1 Four-probe method

The four-probe method is the most commonly used method for PC measurement. The electrodes (probes) are made of noble metals, usually, platinum or gold. In this method, four electrodes are aligned and placed parallel to the plane of the membrane to measure the in-plane conductivity (Figure 4.5) or perpendicular to the membrane to measure the through-plane PC. It is more traditional and easier to find an equipment supplier for the in-plane design. Another advantage of the four-probe method over the two-probe method is that, in the former, the results obtained for PC are more accurate. This is because there is a resistance between the electrode and the membrane surface, which is more irrelevant in this method [48, 49]. In this method, a current is applied in external probes whereas a voltage is measured between the two inner electrodes.

All the *in situ* PC tests measure the resistance of the membrane and not the conductivity. To obtain the PC, the relationship shown in Equation 4.9 must be used:

$$PC = \frac{L}{R \times A} \tag{4.9}$$

where 'L' is the distance between the inner probes; 'R' is the resistance measured in the test; and 'A' is the area of the membrane cross-section.

Compared with impedance, this test offers some advantages, such as the convenience of reproducing the fuel cell environment, quicker testing time, and study of the membrane response to different ranges of RH, temperature and fuels. However, in the impedance test, it is easier to measure the through-plane conductivity and more detailed information such as activation energies (EA) can be obtained.

### 4.3.1.2 Two-probe method

Unlike the four-probe method, in the two-probe method it is easier to obtain the through-plane conductivity. This is one of the biggest advantages of the two-probe method with respect to the four-probe method. However, the contact resistance between the electrode and membrane can compromise the final conductivity values obtained. One popular design of the electrodes is the 'cross shape', where one electrode is in the horizontal plane on one surface of the membrane, and the other is in the vertical plane on the other membrane surface [50].

### 4.3.2 Current–voltage polarisation studies

*In situ* polarisation curves are the most important results defining the PEMFC performance. These curves demonstrate that the behaviour of the fuel cell is in a specific range of current and voltage. However, in a polarisation curve, also called the current × voltage (I–V) curve, it is difficult to identify which losses are due to the membrane or to other components. Figure 4.6 shows a typical polarisation curve.

In Figure 4.6, the line 'a' is defined as the theoretical value of the voltage, which depends on the type of fuel and OT of the PEMFC. For a PEMFC operating <100 °C with hydrogen as fuel, this value is 1.2 V. The relationship between the OCV – top line of the region 'b' – and this theoretical value is known as the 'efficiency' of the fuel cell. The difference between the theoretical value and the OCV are due to fuel crossover and internal currents.

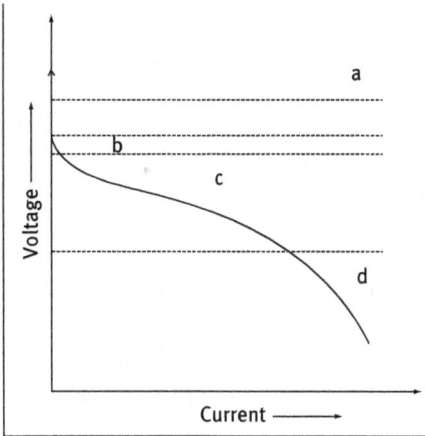

**Figure 4.6:** Typical polarisation curve for a fuel cell.

After the OCV, there is the region 'b' which represents the activation losses in the voltage. These losses are related to fuel redox reactions and the 'sluggishness' of these reactions. The electrodes and catalysts are mostly responsible for these activation losses.

The losses in region 'c' are caused by ohmic losses. These losses, which are linear, fit in Equation 4.10:

$$V = i \times R \tag{4.10}$$

where 'V' is the voltage, 'i' is the current and 'R' is the resistance. The latter, which causes the losses, is a combination of the resistances of all materials and interfaces of the MEA and especially of the resistance of the electrolyte (membrane). The last region, 'd', represents the losses due to mass transport. These losses are due to changes in the concentration of fuel reactions in both electrodes.

From the polarisation curve, the power curves can be obtained. The typical power curve is presented *versus* the current density. The I–V test is a simple test in which the fuel cell and MEA performances are evaluated. However, many parameters must be considered, some of which are discussed below:

- Temperature is probably the first parameter to define in a PEMFC operation. The temperature should be defined before the MEA preparation because the temperature will determine the choice of the membrane [51]. The typical values of OT are 80–100 °C (low temperature), 100–120 °C (IT) and 140–220 °C (HT).
- Fuel: the type of fuel is essential to the membrane choice. Methanol, hydrogen or other types of fuels have very different behaviours. The choice of air or pure oxygen also makes a difference. For example, membranes for DMFC are thicker than those for hydrogen PEMFC due to crossover issues.

- Stoichiometry fuel ratio is the ratio of the real mass flow to the mass consumption rate. The stoichiometry fuel ratio changes between fuel choices. In a fuel cell with $H_2$ and air, the typical values are 1.5/2 [52].
- Gas humidification depends on many factors, for example, the choice of membrane, OT, and pressure. However, even the smallest difference in the humidification level changes the PC and, consequently, the fuel cell performance [53, 54]. Thus, gas humidification is one of the more important parameters from the membrane point-of-view. For Nafion®-like membranes and low-temperature operation, the humidification level to achieve the best performance usually is 60–85%.
- Gas flow is very important to determine the performance of the fuel cell, but also depends on the equipment itself. There is no standard value for this parameter.
- Backpressure determines the state of water (liquid, vapour or mixed). Besides, it affects many other factors, such as membrane conductivity, thermodynamic voltage and mass transfer properties [55].

### 4.3.3 Electrochemical impedance spectroscopy

Electrochemical impedance spectroscopy (EIS) is a powerful method used for studying electrochemical systems such as fuel cells. The system under study is perturbed using a small alternating current (AC) signal and the elicited response of the electrochemical system is then studied in the form of Nyquist plots (real resistance *versus* imaginary resistance) and Bode plots which are plotted with the log of frequency of the AC signal (log $\omega$) on the x-axis and log of absolute impedance (log|Z|) on the y-axis. This is based on the principle that the electrochemical processes taking place at any electrode–electrolyte interface can be interpreted in the form of a representative equivalent electrical circuit with the various electrical elements such as resistors and capacitors. Impedance of an electrical circuit is defined as the effective resistance to an AC signal arising due to combination of the resistors, capacitors and inductors. Using EIS, data-fitted equivalent circuits can be simulated and are used to provide valuable insights about the chemical processes and interactions between the components and effects of specific parameters in the fuel cell. EIS can also be used as a diagnostic tool to identify problematic components/interfaces in an operating fuel cell.

One caution to keep in mind is to ensure the system is in a steady state before EIS is carried out. Unstable systems will very likely result in inaccurate and non-repeatable results in EIS studies. The time required to reach a steady state can vary depending on the system being studied. Gao and co-workers reported that their fuel cell had to be operated for 20 h at 0.6 V before a steady state was reached, Gode and co-workers let their system run for 1 h before carrying out impedance measurements, whereas Pickup and co-workers reported running of an $H_2/O_2$ fuel cell for 30 min at 0.5 V before recording impedance data [56–58]. Another important aspect for EIS measurements is the linearity of the system and the signal amplitude. Ideally, a very low-amplitude AC signal just enough to elicit a linear response (i.e., change in current is proportional to a change in voltage) from the cell is used. The shift between one steady state to another due to the AC-signal perturbation ($\leq 10$ mV) is utilised to derive a mathematical analysis of the electrochemical system under study. The shift between the steady states and rate at which it occurs depends on reaction rate constants, kinetic parameters, charge-transfer resistance, double-layer capacitance, and diffusion coefficients. However, most real systems are not linear. Consequently, a linear approximation or quasi-linear response is measured. Hence, in reality the study involves a trade-off between the signal/noise ratio and the linearity of the response generated by the system [59].

A typical PFSA membrane (which is the most studied) with its hydrophobic backbone and hydrophilic clusters can be treated as a combination of a resistor (R) and capacitance (C) arranged in parallel which can be measured using EIS, and is represented by a semi-circle in the Nyquist plot. The two- and four-probe methods used to study the membrane conductivity can be combined with EIS for more in-depth study of membrane-conductivity behaviour. Xie and co-workers carried out a comparative study on the in-plane conductivity of membranes with two- and four-probe studies combined with EIS on Nafion® 115. They revealed a typical Nyquist plot consisting of a semi-circle followed by a 45° line (for two-probe) and a semi-circle with a geometrically dependent low-frequency feature (for four-probe). The low-frequency intercept of the semi-circle on the x-axis (representing real impedance) was used to determine the membrane resistance [60].

Through-plane conductivity measurements using a four-probe setup can also be studied using a single fuel cell or a fuel cell stack in great detail by varying and controlling parameters such as RH and temperature to systematically study the variation in membrane conductivity with respect to these parameters over time.

Most studies investigate the MEA system as a whole in *in situ* studies, treating the membrane as one unit (as a resistor) with ionic conductivity as its main parameter. *In situ* EIS studies provide much more interesting and complex data due to the complexity of the number of components and interfaces involved. When studying a MEA, the anode and cathode are treated as two electric circuits with different RC values. With the membrane as R in the middle the three components together represent a whole circuit (Figure 4.7). The anode reaction is relatively faster, so the representative RC circuit for the anode side of a fuel cell is often ignored. EIS can also be carried out as *ex situ* in combination with other tests, such as ionic conductivity, as discussed above.

In *ex situ* studies using EIS, Bayer and co-workers undertook ionic-conductivity studies using specialised equipment to obtain PC and other behaviour of nanocellulose fibre (21 µm) and nanocellulose crystal (33 µm) membranes (Figure 4.8). They

(a)

(b)

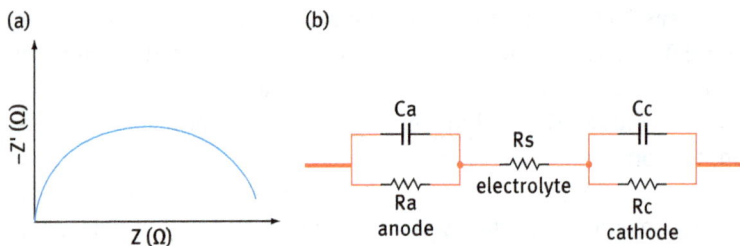

**Figure 4.7:** Typical Nyquist plot with a semi-circle representing the membrane and the equivalent circuit for a simple fuel cell system (Ca: anode capacitance; Cc: cathode capacitance; Ra: anode resistance; Rc: cathode resistance; and Rs: solution resistance).

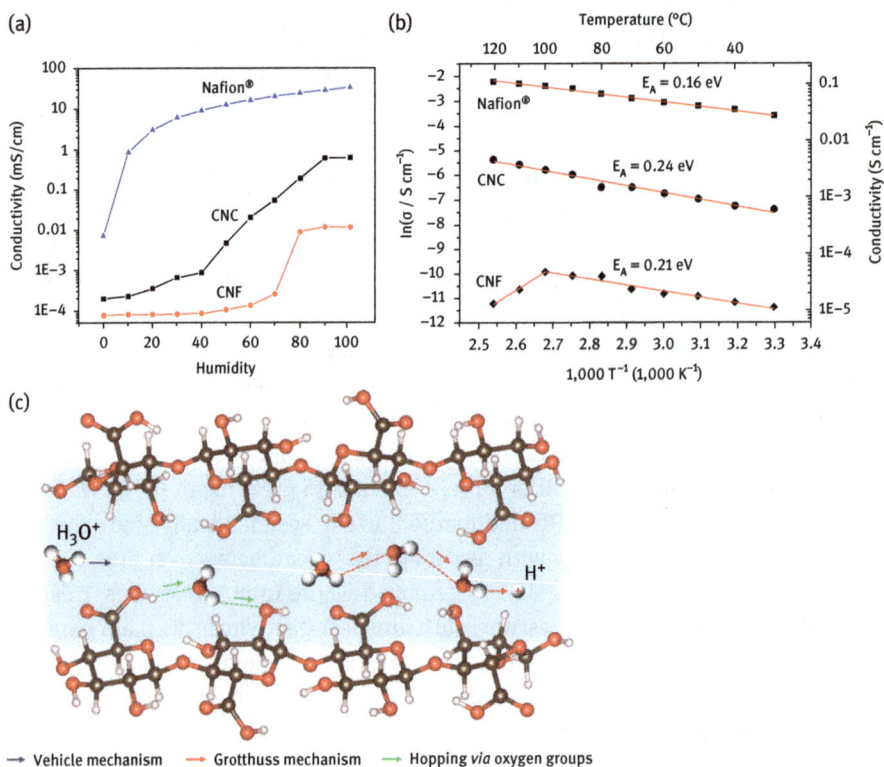

—→ Vehicle mechanism —→ Grotthuss mechanism —→ Hopping *via* oxygen groups

**Figure 4.8:** PC behaviour in nanocellulose. (a) Dependence of PC on RH for CNF, CNC paper, and Nafion® at 30 °C. (b) Arrhenius plot at 100% RH with EA. (c) Schematic of possible proton-conduction mechanisms and pathways in nanocellulose. Reproduced with permission from T. Bayer, B.V. Cunning, R. Selyanchyn, M. Nishihara, S. Fujikawa, K. Sasaki and S.M. Lyth, *Chemistry of Materials*, 2016, **28**, 13, 4805. ©2016, American Chemical Society [15].

investigated ionic conductivity for a range of temperatures and RH using a membrane-testing system (740 MTS, Scribner) coupled with an impedance analyser (Solartron SI 1260). The membranes were measured using an AC amplitude of 10 mV at 30 MHz to 10 Hz. The impedance measurements were carried out at 30 °C with isothermal changes in $100 \leq RH \leq 0\%$. The membranes were pre-treated at 100% RH for 4 h, followed by 1-h pre-treatment at each humidity. To study the influence of OT on the PC, the impedance was measured from 40 to 120 °C at 100% RH. Above 100 °C, the measurement chamber was pressurised at a total pressure of 230 kPa. Finally, the obtained impedance plots were fitted to an equivalent circuit from which the membrane resistance, R ($\Omega$), was determined from the high-frequency intercept.

Their tests revealed that, above 100 °C, the conductivity of CNF dropped to 0.01 mS /cm, which was attributed to reduced WU at these temperatures above the boiling point of water. However, the conductivity of the CNC membrane was found to increase continuously from 0.62 mS/cm at 30 °C to 4.57 mS/cm at 120 °C. The higher conductivity of CNF was ascribed to the acid hydrolysis treatment they were subjected to, which introduced sulfonic acid groups to the structure, thereby increasing the number of protons available for conduction. The authors also determined the proton transport EA from the slopes of the Arrhenius plot, and revealed similar EA for CNC and CNF. The EA of 0.21 and 0.24 eV was slightly higher than that of Nafion® (0.16 eV) at high humidity, where the Grotthuss mechanism dominates. Based on this finding, the authors speculated on the possibility of a 'nanoconfinement' effect in the channels within the cellulose fibres, between microfibrils, resulting in an increase in EA and the prospect of proton hopping along the oxygen groups of the cellulose at low humidity [15].

### 4.3.3.1 Use of electrochemical impedance spectrometry to study the effects of membrane thickness

An EIS study was carried out to investigate the effect of membrane thickness under different operating conditions. A study by Freire and Gonzalez explored four types of Nafion® membranes [117 (175 μm), 115 (125 μm), 1135 (80 μm) and 112 (50 μm)] in an $H_2/O_2$ fuel cell with an active area of 1 cm² operating under atmospheric conditions [61]. The authors used a Solartron SI 1250 frequency analyser coupled with a Solartron SI 1286 electrochemical interface in potentiostatic mode at an input AC signal of 10 mV. The obtained plots were corrected for ohmic drop in the single cell using high-frequency resistance. Their investigations revealed that cells operating (at RT and fully humidified gases) with Nafion® 117 and 112 were of particular interest. For these systems, at higher cell potentials (0.9 and 0.8 V), a 45° branch was seen in the impedance spectra at the high end of the frequency range. This line behaviour in the Nyquist plot has been observed previously and is associated with the Warburg impedance component in the equivalent circuit. This high-frequency region is dominated by proton transport and double-layer charging behaviour because the Faradaic responses are non-significant at these frequencies. Consequently, within the operating fuel cell, this behaviour is attributed to limited proton transport in the electrode due to the thin film of the electrolyte. Springer and co-workers had also attributed this to the coupling of distributed ionic resistance and capacitance behaviour as a consequence of the limited PC in electrode [62]. Freire and Gonzalez also observed a second semi-circle for Nafion® 117 in the lower end of the frequency range, the size of which appeared to increase with a decrease in electrode potential. This feature was attributed to the ineffective mass transport of the reaction gases on the electrode surface, which resulted from flooding at the cathode and was not observed for Nafion® 112 at low-electrode potentials. Their studies at other temperatures (80 and 85 °C for $O_2$; 95 °C for $H_2$) with humidified gases also demonstrated that thicker membrane resulted in cathode flooding as a consequence of inefficient back diffusion through the thickness of the membrane. A thicker membrane also demonstrated a linear dependence of high-frequency resistance on membrane thickness, suggesting that thicker membranes no longer behave as pure resistors at high current densities because of the capacitive effect resulting from ineffective back diffusion. They also concluded that thinner membranes allowed better water management due to effective back diffusion and were also more tolerant to variation in humidity and temperature conditions, and current density.

### 4.3.4 Other *in situ* tests

*In situ* studies such as open circuit (often undertaken in combination with Fenton's test) and load cycling (in which RH conditions have a significant role) are covered in more detail in Chapter 5. Thermal stability for MEA is tested by carrying out F/T cycling and a sub-zero start-up (also discussed in Chapter 5).

# References

[1]   K. Ketpang, K. Oh, S. Lim and S. Shanmugam, *Journal of Power Sources*, 2016, **329**, 441.
[2]   Y. Kim, K. Ketpang, S. Jaritphun, J.S. Park and S.Shanmugam, *Journal of Materials Chemistry A*, 2015, **3**, 15, 8148.
[3]   J.T. Hinatsu, M. Mizuhata and H. Takenaka, *Journal of the Electrochemical Society*, 1994, **141**, 6, 1493.
[4]   T.A. Zawodzinski, T.E. Springer, J. Davey, R. Jestel, C. Lopez, J.Valerio and S. Gottesfeld, *Journal of the Electrochemical Society*, 1993, **140**, 7, 1981.
[5]   H. Beydaghi, M. Javanbakht and E. Kowsari, *Industrial & Engineering Chemistry Research*, 2014, **53**, 43, 16621.
[6]   T. Sata in *Ion Exchange Membranes: Properties, Characterization and Microstructure of Ion Exchange Membranes*, Ed., T. Sata, Royal Society of Chemistry, London, UK, 2004, p.89.
[7]   S.J. Lue, Y. Pai, C. Shih, M. Wu and S. Lai, *Journal of Membrane Science*, 2015, **493**, 212.

[8]    S. Yılmaztürk, H. Deligöz, M. Yılmazoğlu, H. Damyan, F. Öksüzömer, S.N. Koc, A. Durmuş and M.A. Gürkaynak, *Journal of Power Sources*, 2010, **195**, 3, 703.

[9]    C. Lin, W. Lien, Y. Wang, H. Shiu and C. Lee, *Journal of Power Sources*, 2012, **200**, 1.

[10]   M. Tohidian, S.R. Ghaffarian, M. Nouri, E. Jaafarnia and A.H. Haghighi, *Journal of Macromolecular Science: Part B*, 2015, **54**, 1, 17.

[11]   K. Peinemann and S.P. Nunes in *Membrane Technology*, Eds., K. Peinemann and S.P. Nunes, Volume 2, John Wiley & Sons, New York, NY, USA, 2008.

[12]   S.J. Peighambardoust, S. Rowshanzamir and M. Amjadi in *Review of the Proton Exchange Membranes for Fuel Cell Applications*, Elsevier Ltd, Amsterdam, The Netherlands, 2010, p.9349.

[13]   S. Feng and G.A. Voth, *The Journal of Physical Chemistry B*, 2011, **115**, 19, 5903.

[14]   T. Bayer, B. V. Cunning, R. Selyanchyn, T. Daio, M. Nishihara, S. Fujikawa, K. Sasaki and S.M. Lyth, *Journal of Membrane Science*, 2016, **508**, 51.

[15]   T. Bayer, B.V. Cunning, R. Selyanchyn, M. Nishihara, S. Fujikawa, K. Sasaki and S.M. Lyth, *Chemistry of Materials*, 2016, **28**, 13, 4805.

[16]   J. Yan, X. Huang, H.D. Moore, C. Wang and M.A. Hickner, *International Journal of Hydrogen Energy*, 2012, **37**, 7, 6153.

[17]   B.P. Tripathi and V.K. Shahi, *The Journal of Physical Chemistry B*, 2008, **112**, 49, 15678.

[18]   Q. Wang, W. Guoxiong, L. Xing, C. Chunhuan, L. Zhiqiang and S. Gongquan, *International Journal of Electrochemical Science*, 2015, **10**, 2939.

[19]   V. Parthiban, S. Akula, G.S. Peera, N. Islam and A.K. Sahu, *Energy & Fuels*, 2016, **30**, 1, 725.

[20]   M.E. Lindsey and M.A. Tarr, *Chemosphere*, 2000, **41**, 3, 409

[21]   C. Walling, *Accounts of Chemical Research*, 1975, **8**, 4, 125.

[22]   C. Walling, *Accounts of Chemical Research*, 1998, **31**, 4, 155

[23]   M. Zatoń, J. Rozière and D. Jones, *Sustainable Energy & Fuels*, 2017, **1**, 3, 409.

[24]   J. Healy, C. Hayden, T. Xie, K. Olson, R. Waldo, M. Brundage,H. Gasteiger and J. Abbott, *Fuel Cells*, 2005, **5**, 2, 302.

[25]   H. Tang, M. Pan, F. Wang, P.K. Shen and S.P. Jiang, *The Journal of Physical Chemistry B*, 2007, **111**, 30, 8684.

[26]   T. Kinumoto, M. Inaba, Y. Nakayama, K. Ogata, R. Umebayashi, A. Tasaka, Y. Iriyama, T. Abe and Z. Ogumi, *Journal of Power Sources*, 2006, **158**, 2, 1222.

[27]   M. Aoki, H. Uchida and M. Watanabe, *Electrochemistry Communications*, 2005, **7**, 12, 1434.

[28]   Z. Chang, H. Yan, J. Tian, H. Pan and H. Pu, *Polymer Degradation and Stability*, 2017, **138**, 98.

[29]   H. Tang, S. Peikang, S.P. Jiang, F. Wang and M. Pan, *Journal of Power Sources*, 2007, **170**, 1, 85.

[30]   J.M. Alsheheri, H. Ghassemi and D.A. Schiraldi, *Journal of Power Sources*, 2014, **267**, 316.

[31]   M.T. Salleh, J. Jaafar, M.A. Mohamed, M.N.A.M. Norddin,A.F. Ismail, M.H.D. Othman, M.A. Rahman, N. Yusof, F. Aziz and W.N.W. Salleh, *Polymer Degradation and Stability*, 2017, **137**, 83.

[32]   D. Liu, J. Li, J. Ni, X. Xiang, B. Liu and L. Wang, *RSC Advances*, 2017, **7**, 15, 8994.

[33]   K.H. Wong and E. Kjeang, *ChemSusChem*, 2015, **8**, 6, 1072.

[34]   A. Reyna-Valencia, S. Kaliaguine and M. Bousmina, *Journal of Applied Polymer Science*, 2005, **98**, 6, 2380.

[35]   Y. Tang, A.M. Karlsson, M.H. Santare, M. Gilbert, S. Cleghorn and W.B. Johnson, *Materials Science and Engineering: A*, 2006, **425**, 1, 297.

[36]   A. Kusoglu, A.M. Karlsson, M.H. Santare, S. Cleghorn and W.B. Johnson, *Journal of Power Sources*, 2007, **170**, 2, 345.

[37]   D.M. Yu, S. Yoon, T. Kim, J.Y. Lee, J. Lee and Y.T. Hong, *Journal of Membrane Science*, 2013, **446**, 212.

[38] *Aquivion® PFSA*, Solvay, Brussels, Belgium. http://www.solvay.com/en/markets-and-prod ucts/featured-products/Aquivion.html [last accessed June 2017]
[39] Y. Tang, A. Kusoglu, A.M. Karlsson, M.H. Santare,S. Cleghorn and W.B. Johnson, *Journal of Power Sources*, 2008, **175**, 2, 817.
[40] Y. Cai, Z. Yue and S. Xu, *Journal of Applied Polymer Science*, 2017, **134**, 25,
[41] T. Weissbach, T.J. Peckham and S. Holdcroft, *Journal of Membrane Science*, 2016, **498**, 94.
[42] W. Liu, K. Ruth and G. Rusch, *Journal of New Materials for Electrochemical Systems*, 2001, **4**, 4, 227.
[43] Y. Li, D.A. Dillard, S.W. Case, M.W. Ellis, Y. Lai,C.S. Gittleman and D.P. Miller, *Journal of Power Sources*, 2009, **194**, 2, 873.
[44] K. Patankar, D.A. Dillard, S.W. Case, M.W. Ellis, Y. Li, Y. Lai, M.K. Budinski and C.S. Gittleman, *Journal of Polymer Science, Part B: Polymer Physics*, 2010, **48**, 3, 333.
[45] R.C. McDonald, C.K. Mittelsteadt and E.L. Thompson, *Fuel Cells*, 2004, **4**, 3, 208.
[46] X. Chen, L. Yan, Z. Wang and D. Liu, *Journal of Power Sources*, 2011, **196**, 5, 2644.
[47] C.M. Branco, S. Sharma, M.M. De Camargo Forte and R. Steinberger-Wilckens, *Journal of Power Sources*, 2016, **316**, 139.
[48] C. Prapainainar and S.M. Holmes in *Sustainability in Energy and Buildings: Research Advances*, Mediterranean Green Energy Forum, Future Technology Press, Shoreham-by-Sea, UK, 2013, p.31.
[49] C.H. Lee, H.B. Park, Y.M. Lee and R.D. Lee, *Industrial & Engineering Chemistry Research*, 2005, **44**, 20, 7617.
[50] A. Kishi and M. Umeda, *Electrochemistry*, 2007, **75**, 2, 130.
[51] K. Scott, C. Xu and X. Wu, *Wiley Interdisciplinary Reviews: Energy and Environment*, 2014, **3**, 1, 24.
[52] L. Wang, J. Kang, J. Nam, J. Suhr, A.K. Prasad and S.G. Advani, *ECS Electrochemistry Letters*, 2015, **4**, 1, F1.
[53] C. Sun, K.L. More, G.M. Veith and T.A. Zawodzinski, *Journal of the Electrochemical Society*, 2013, **160**, 9, F1000.
[54] R.S. Fu, J.S. Preston, U. Pasaogullari, T. Shiomi, S. Miyazaki, Y. Tabuchi, D.S. Hussey and D.L. Jacobson, *Journal of the Electrochemical Society*, 2011, **158**, 3, B303.
[55] J. Zhang, C. Song, J. Zhang, R. Baker and L. Zhang, *Journal of Electroanalytical Chemistry*, 2013, **688**, 130.
[56] Q. Guo, M. Cayetano, Y. Tsou, E.S. De Castro and R.E. White, *Journal of the Electrochemical Society*, 2003, **150**, 11, A1440.
[57] E.B. Easton and P.G. Pickup, *Electrochimica Acta*, 2005, **50**, 12, 2469.
[58] P. Gode, F. Jaouen, G. Lindbergh, A. Lundblad and G. Sundholm, *Electrochimica Acta*, 2003, **48**, 28, 4175.
[59] X. Yuan, C. Song, H. Wang and J. Zhang in *Electrochemical Impedance Spectroscopy in PEM Fuel Cells: Fundamentals and Applications*, Springer-Verlag, Heidelberg, Germany, 2009.
[60] Z. Xie, C. Song, B. Andreaus, T. Navessin, Z. Shi, J. Zhang and S. Holdcroft, *Journal of the Electrochemical Society*, 2006, **153**, 10, E173.
[61] T.J. Freire and E.R. Gonzalez, *Journal of Electroanalytical Chemistry*, 2001, **503**, 1, 57.
[62] T. Springer, T. Zawodzinski, M. Wilson and S. Gottesfeld, *Journal of the Electrochemical Society*, 1996, **143**, 2, 587.

# 5 Membranes in single-cell proton-exchange membrane fuel cells and stacks

## 5.1 Introduction

*In situ* test obviously provide a very real analysis as no *ex situ* test can create the dynamic and complex environment which the membrane must face inside a fuel cell. However, *ex situ* tests will continue to have limitations because each method can usually investigate only specific characteristics with limited variables. Moreover, they cannot simulate the exact environment inside the fuel cell so the precision of these tests is also restricted. For example, water uptake, ion-exchange capacity (IEC) and proton conductivity (PC) studies often lead to contradictory observations about membrane behaviour.

The challenges faced by the electrolyte membrane as a consequence of the alternations in other components and parts of an operating fuel cell can be studied even more closely by combining various specialised physical characterisation methods with *in situ* test setups. The variety of accelerated stress test (AST) procedures used in *in situ* studies (such as accelerated mechanical stress tests (AMST), freeze/thaw (F/T) cycling) evolve an understanding of the changes faced by a membrane and the forces incumbent upon it during the lifetime of a cell. Although only a limited number of *in situ* AST studies have been conducted, which could be due to the expensive nature of such prolonged testing using specialised equipment, they are imperative to the commercialisation of fuel cell systems. AST studies on the polymer electrolyte membrane (PEM) (as well as other components) provide essential information on the failure modes of the system, the analysis of which is crucial given the highly integrated proton-exchange membrane fuel cell (PEMFC) system [1]. AST protocols are designed to assess the durability and diagnose potential mitigation strategies at judicious cost and time by exacerbating the rate of chemical and mechanical degradation. To minimise experimental time and enhance the efficacy of the AST studies a number of unfavourable and detrimental operating conditions are utilised: relative humidity (RH) cycling, F/T cycling, current density, air starvation and open circuit voltage (OCV) operation [2–7].

The first part of this chapter explores the latest developments towards understanding membrane behaviour (including water diffusion, transport and management, proton conduction as well as gas permeation) under various standard and extreme conditions. The use of single-cell PEMFC systems and/or specialised cells/equipment in combination with high-end visualisation (such as transmission electron microscopy (TEM) and cryogenic tomography) methods and other advanced material characterisation methods [vibrational spectroscopy, nuclear magnetic resonance (NMR)] in recent years have led to significant insights towards understanding of perfluorosulfonic acid (PFSA), (especially Nafion$^{®}$) as well as other

https://doi.org/10.1515/9783110647327-005

membranes. Following that discussion, recent AST studies focussing on single cell as well as small stack data on Nafion® membrane behaviour (durability and degradation) along with studies exploring Nafion® performances under extreme conditions and specialist applications will be discussed. The last part of this chapter delves into *in situ* studies on other PFSA–Nafion® composite and non-PFSA membranes for PEMFC and direct methanol fuel cell (DMFC) applications. *In situ* studies on non-Nafion® membranes have gained momentum only recently because these membranes have slowly progressed towards the next level of assessment after extensive *ex situ* studies over the last decade.

## 5.2 Combining *in situ* and physical characterisation methods: Water behaviour inside a membrane

### 5.2.1 Diffusion of water and polymer reorganisation in Nafion®

Despite the extensive knowledge base accrued over the years on PFSA membranes, our understanding of them is still evolving. This section discusses the phenomenon of water diffusion, distribution and transport in a membrane in an operating fuel cell. However, before proceeding, emphasis must be placed on the recent new developments achieved through *ex situ* studies with the help of specialised physical characterisation using NMR and small-angle neutron scattering (SANS).

The nanostructure of PFSA is frequently studied using SANS. This technique uses a well-defined correlation peak (ionomer peak), which is highly sensitive to water uptake, and identifies the mean separation distances between hydrophobic or hydrophilic domains (i.e., d-spacing). It is known that membranes swell 5-fold more at the nanoscale than at the macroscale [8]. Reports have pointed out that to obtain the mutual diffusion coefficient ($D_m$) the hydration kinetics data for membranes (generally obtained by placing the membrane inside humidified chambers at controlled vapour pressure) are typically fitted using the solution of second Fick's law. However, most experimental cell setups do not allow for a constant water concentration at the membrane surface (which is a requirement for using Fick's model) and instead only have a low flow of humidified gas. This results in inaccurate $D_m$ values spread over several orders of magnitude and much lower than those of the water self-diffusion coefficient ($D_s$) [9–12]. Thus, water uptake in such experiments must be controlled by the interfacial transport resistance, which biases the $D_m$ measurements [10]. Taking this to the next step, Fumagalli and co-workers carried out SANS studies on Nafion® 117 (equivalent weight = 1,100 g/eq, 175 μm) using a specially designed stopped-flow device to rapidly inject water into the membrane inside the cell, thus overcoming the influence of interfacial effects (contacting the membrane with liquid water forces a constant boundary condition during the sorption process). A 1 × 2 cm$^2$ Nafion® membrane was inserted in a 1 mm-thick stopped-flow cell made of quartz. The kinetics experiments were done at 5, 15, 35, and 50 °C. For imaging, the authors used a reduced time resolution of 200 ms for the first 300 images, then 60 images taken every 10 s, and finally 40 images/ min. Static measurements carried out at 25 °C were done on the same acidified membrane sample after 7 and 14 h of immersion in water (at 25 or 80 °C). The long-term behaviour was probed on samples (acidified or as-received) previously immersed in water for days, weeks, months, and years. The authors also ensured that all data presented in their work was taken in the absence of air bubbles by controlling the presence of bubbles using a camera and immediately detecting abrupt variations in the incoherent background level. Details of the SANS setup were an incoming wavelength '$\lambda$' = 5 Å and sample-to-detector distance = 2.5 m such that scattering wave vectors 'q' [q = $4\pi$/ $\lambda \sin(\theta)$, where $\theta$ is the scattering angle] were in the range q = 0.02–0.48 Å$^{-1}$. This q-range was selected such that the ionomer peak position fell within this q-window at all hydration states. The constant incoherent background due to water molecules at q > 0.25 Å$^{-1}$ enabled precise control of water content during experiments. These time-resolved experiments led to four interesting observations. The first was nanoscopic swelling – within the first 30 s, the d-spacing increased dramatically from 33 to 43 Å (similar to those reported by Kusoglu and co-workers [13]) followed by a slow (yet continuous) increase in d-spacing value. The $D_m$ values thus obtained were much closer to those of $D_s$. Second, the integrated intensity values studied for pre-hydrated and pre-dried membranes were identical after the first 25 s, suggesting that a 'master' kinetic law, independent from the initial state of the membrane, was quickly recovered after the initial fast sorption process. Third, no swelling equilibrium was reached even after years because long-term nanoscopic swelling behaviour was identified based on an inter-domain mean separation distance which, according to log(t) dependence, increased continuously over almost seven decades. Within 20 s of water injection, the size of ionic domains increased by 25%, 37% after 1 day, 39% after 2 weeks and 43% (i.e., d = 47.8 Å) in 6 years (Figure 5.1). Fourth, long-term static measurements undertaken on as-received and re-acidified membranes equilibrated at 80 ° C showed that absolute d-spacing values were affected in both cases but the as-received membrane swelling was more limited as compared with the re-acidified sample (which was attributed to multivalent cation pollutants exchanging some acidic sites) and higher d values were found at increased temperatures based on a report by Rollet and co-workers [8, 14].

Based on these observations the authors proposed the three-step sorption process described below:
- Step 1: Ionisation and solvation of ionic headgroups – severe spatial confinement and strong interactions of ionic species with fluid molecules stimulates a slow and low-rate penetration of fluid/water molecules.
- Step 2: Diffusion of water molecules – fast and substantial nanoscale swelling of the membrane occurs as thermally-activated Fickian diffusion of molecules takes place inside the well-connected network of ionic channels. The water

(a)

(b)

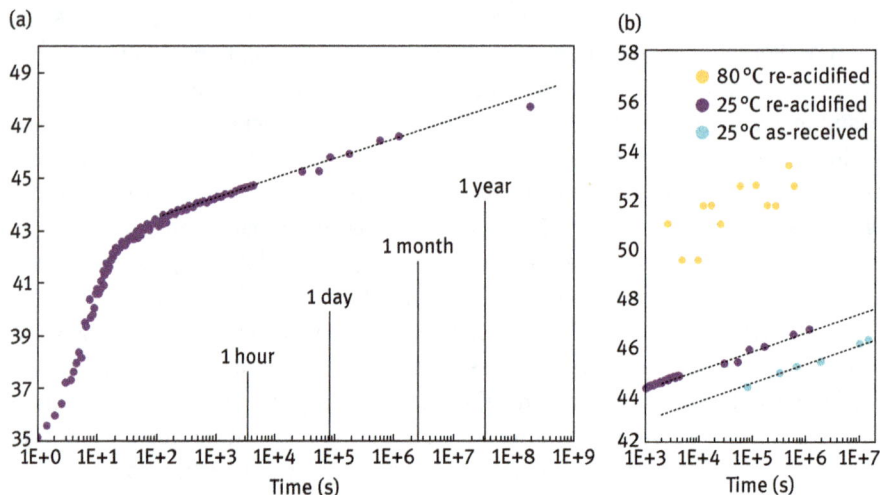

**Figure 5.1:** a) Trend of d-spacing as a function of the logarithm of time, log(t), measured at 25 °C for a Nafion® 117 equilibrated previously at ambient humidity and b) d-spacing variations in the as-received and pre-treated Nafion® 117. Reproduced with permission from M. Fumagalli, S. Lyonnard, G. Prajapati, Q. Berrod, L. Porcar, A. Guillermo and G. Gebel, *The Journal of Physical Chemistry B*, 2015, **119**, 23, 7068. ©2015,American Chemical Society [8].

concentration gradients formed between the membrane surface and core produces mutually diffusive coefficients, which are compatible with water $D^s$ at the microscale and which govern this nanoscale swelling process.

– Step 3: Reorganisation of the polymer matrix – in this step, surplus water molecules are accommodated continuously within the ionic phase over months and years. This suggests that the polymer behaves like a solid-like polymer above the yield point rather than a viscoelastic polymer, which would deform over a long term. The polymer reorganisation allows a restricted, but efficient and continuous incorporation, of water molecules within the membrane over several years [8].

The heat treatment, temperature, and drying process to which the membrane may be subjected to during the pre-treatment or membrane electrode assembly (MEA) preparation also affect its sorption capacity. Maldonado and co-workers investigated Nafion® 115 and Nafion® NR-212 to study the effect of drying temperature on the PC, water content and the $D_s$ using NMR and pulsed-field gradient nuclear magnetic resonance (PFGNMR). Their work revealed fascinating insights and concluded that when the membrane samples are heat treated to a moderate temperature (between 60 and 100 °C), some residual water remains strongly attached to the polymer structure, and that the amount of residual water can be determined using NMR in terms of $\lambda$ (i.e., number of water molecules per ionic site) ($\lambda = 1.5 \pm 0.5$ for Nafion® 115). They also found that the sorption capacity suffered in heat-treated samples during drying as a consequence of shrinkage in the polymer structure. While this effect was evident in liquid as well as vapour phases and increased with the drying temperature, its effect decreased with increasing measuring temperature. Subsequently, the values of PC ('σ') and $D_s$ were also reduced due to this loss of sorption capacity. However, when the measuring temperature increased, the effects of temperature dominated over the effects of hydration. That study pointed out the prominence of the hygro-thermomechanical properties of Nafion® when evaluating its properties [15].

Kaddaouri and co-workers also carried out PFGNMR studies on Nafion® 1110 (254 μm) and investigated the quantitative effects of static normal stress on its water sorption and diffusion properties when equilibrated in water vapour. They devised an NMR-compatible compression setup (Figure 5.2) in which the membranes were subjected to stress at 0–12 MPa at room temperature (RT) (24 ± 2 °C) under variable RH (15–98%). This setup also enabled the simultaneous determination of the $D_s$ along three spatial orthogonal directions. It must be stressed that in all these reported studies the membranes were thoroughly pre-treated and cleaned of any impurities that could interfere with the results. The procedures varied with the researchers' protocols so it is not possible to detail them in every literature report discussed in this book. The comprehensive and meticulous work by Kaddouri and co-workers revealed three main findings. First, the external applied force can be balanced by the internal elastic pressure in a membrane equilibrated at high RH (>90%), which results in a substantial decrease in its water content. This effect was observed for the low stress (0–3 MPa) the MEA was subjected to in the form of clamping pressure in a working PEMFC. The effect of compressive stress on the water content is shown in Figure 5.3. Second, the elastic and osmotic pressures inside the membrane become very high at RH = 0–85%, consequently the compression had limited effects on water content. Third, the diffusion measurements by Kaddouri and co-workers demonstrated that water mobility was also affected in the stressed membrane as stress-stimulated micrometric-scale structural modifications or reorganisations, which was evident from the decrease in the values of $D_s$. These effects could affect the PC inside the PEM [16].

## 5.2.2 Water transport through the membrane

Membrane hydration defines the performance and durability of a PEMFC. An improperly hydrated membrane exhibits undesirably high ionic resistance and may suffer from irreversible damages in extreme circumstances. Membrane hydration inside the operating PEMFC is governed by water transport, which is mainly driven by 'electro-osmotic drag', back diffusion and convection. The net water transport inside the operating fuel cell, thus, is affected by the conditions of inlet gases (level of humidification, gas pressure and stoichiometry) and other operating conditions

(a)　　　　　　　　　　　　　　　　　　　　　(b)

| 1 - Flat head screw (amagnetic) | 2 - Chamber cap | 3 - Spacer |
| 4 - Cover | 5 - Piston with o-ring seal | 6 - Chamber |
| 7 - Clamping nut (amagnetic) | 8 - Porous filter disc (quartz) | 9 - Membrane |
| 10 - 2D NMR coil | | |

(c)　　　　　　　　　　　　　　　　　　　　　(d)

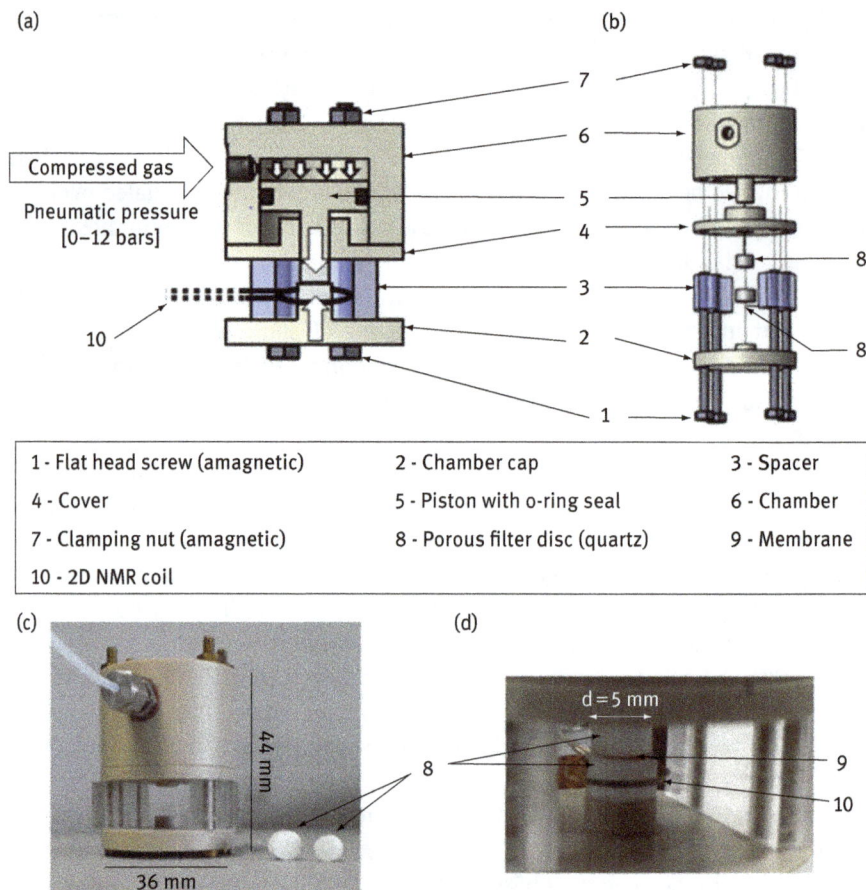

**Figure 5.2:** Schematics and pictures of the compression NMR setup: (a) operation scheme with a two-dimensional (2D) NMR coil, (b) exploded view, (c) picture of the compression setup and two porous discs, and (d) picture of the compressive zone with porous discs, the membrane (d = 5 mm), and 2D NMR coil. Reproduced with permission from A. El Kaddouri, J. Perrin, T.Colinart, C. Moyne, S. Leclerc, L. Guendouz and O. Lottin, *Macromolecules*, 2016, **49**, 19, 7296. ©2016, American Chemical Society [16].

(current density, current load, temperature) of the cell. Earlier modelling experiments have established the need for humidifying the anode gas supply, especially at high current densities ($>1$ A/cm$^2$), where a large fraction of voltage loss is due to membrane ohmic losses and back diffusion cannot retain a hydrated membrane [18]. Modelling studies have also revealed that the fluctuating water requirement inside the fuel cell can be adjusted by regulating the humidification of reactant gases as the current density is varied [19–21]. Motupally and co-workers further identified the role of water activity gradients in the diffusion of water across the membrane and also compared the Fickian diffusion coefficient models proposed

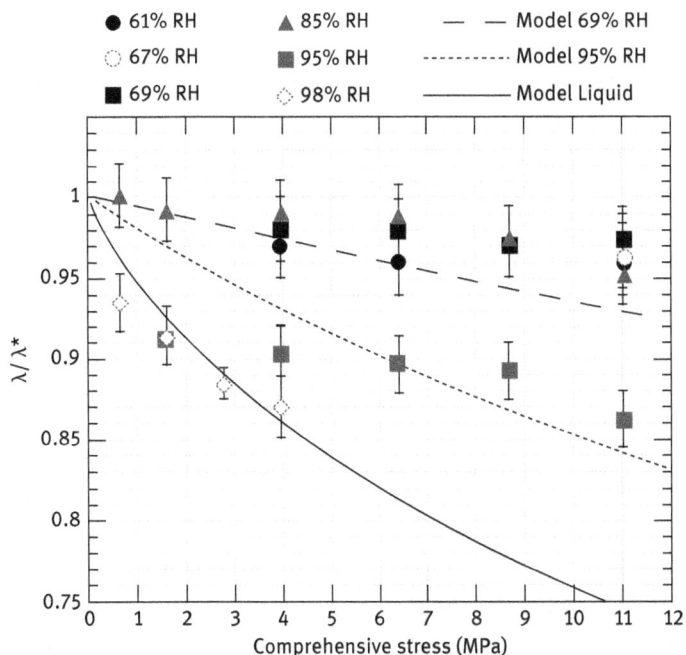

**Figure 5.3:** Effect of a variable compressive stress on the water content for different RH. $\lambda^*$ is the water content of the membrane as measured at equilibrium before compression. Lines represent the simulated results of the effect of stress on water content using the micro/macro-homogenisation method adapted in Nafion®by Colinart and co-workers [17] for RH = 69% and 95%. Reproduced with permission from A. El Kaddouri, J. Perrin, T. Colinart, C. Moyne, S. Leclerc, L. Guendouz and O. Lottin, *Macromolecules*, 2016, **49**, 19, 7296. ©2016, American Chemical Society [16].

earlier by Fuller, Nguyen and co-workers, and Zawodzinski and co-workers [18, 22–24]. Yan and co-workers experimentally determined the net drag coefficient at different humidity conditions. They collected discharged water by condensing and trapping the effluent water from the anode and cathode [25]. Those studies on the effect of feed gas humidity on water transport demonstrated three main features. First, when nearly dry (RH = 10%) air and fully-hydrated $H_2$ were supplied at the cathode and anode, respectively, it was mainly the product (water) which diffused from the cathode–membrane interface to the channel along with the water from the anode delivered by the electro-osmotic drag which humidified the cathode. Second, when the RH of air was increased to 100% (at 80 °C), the water gradient through the membrane increased due to increased water content in the air. This phenomenon led to an increase in back flux, thus decreasing the net drag coefficient. Third, the contribution of anode gas humidification became more important as the current density increased and, as Yan and co-workers observed, the supply of dry hydrogen generated a large-negative drag that could dry out the cathode even in the presence of saturated oxygen [25].

### 5.2.3 Water distribution inside the membrane

Various visualisation studies, earlier using optical photography and later using NMR, neutron imaging, and environmental scanning electron microscopy, have also been employed to understand the build-up, distribution, and removal of water not just within the membrane but in other components as well [26–29]. Lee and co-workers operated a PEMFC of active area of 25 cm$^2$ in dead-end mode to study the water transport on the basis of performance degradation of the PEMFC and visualisation of variation in water accumulation at the anode. Their system consisted of a four-channel serpentine-style flow field machined on a graphite current collector for use on the cathode side; at the anode, a 1 mm-thick gold-coated stainless-steel plate was used as the current collector with flow field penetrated on the plate to allow hydrogen flow along the channel. A transparent, polycarbonate window was utilised to view water accumulation at the anode, which enabled a charge-coupled device camera to capture images every 30 s. Using stoichiometries of 2.5 and 1.5 for air and hydrogen, respectively, and RH maintained at 100%, the cell was operated at 70 °C. Their study evaluated MEA with Nafion® 112 (50 μm) and Gore PRIMERA® 57 series (18 μm) as catalyst-coated membranes (CCM). Lee and co-workers evaluated the effect of anodic, cathodic and bilateral humidification, and reported that the highest rate of voltage reduction was observed in the case of bilateral humidification. Although during anodic humidification no voltage reduction was observed for 60 min, there were intermittent fluctuations. They also observed that the water diffusivity of the Gore® membrane was almost half that of Nafion® because of the reinforced Gore® membrane [30]. While studying the variation in voltage with respect to humidity, it was found that at low humidity the voltage decreased slowly with time, in Nafion®-based and Gore®-based MEA. However, a significantly longer time duration was required to reach the limit of 0.2 V in the case of Nafion® 112. The water flux of Nafion® by diffusion was ≈0.7-fold that of Gore® because its diffusivity was more than 2-fold and the thickness nearly 2.8-fold that of Gore®. The slower diffusion reduced the rate of water build-up, thus delaying the reduction in voltage to reach the limit of 0.2 V. However, at low RH, as the water accumulated at the anode, more of it was used for membrane hydration and, consequently, much more water transporting through membrane seemed to be captured inside the Nafion® membrane, which was 2.8-fold thicker than the Gore® membrane [31].

Vibrational spectroscopic methods have also been explored to understand water behaviour inside membranes [32–34]. These offer advantages such as higher sensitivity towards water vapour and small 'water clusters' over NMR and X-ray scattering methods, which are more suited for liquid–water analyses. Kunimatsu and co-workers studied hydration/dehydration cycles in Nafion® (NR-211) at RT using combined time-resolved attenuated total reflection (ATR)–Fourier–transform infrared (FTIR) spectroscopy. A specially designed ATR cell consisting of a carbon separator with interdigitated-type gas fields was used (Figure 5.4). Aluminium foil

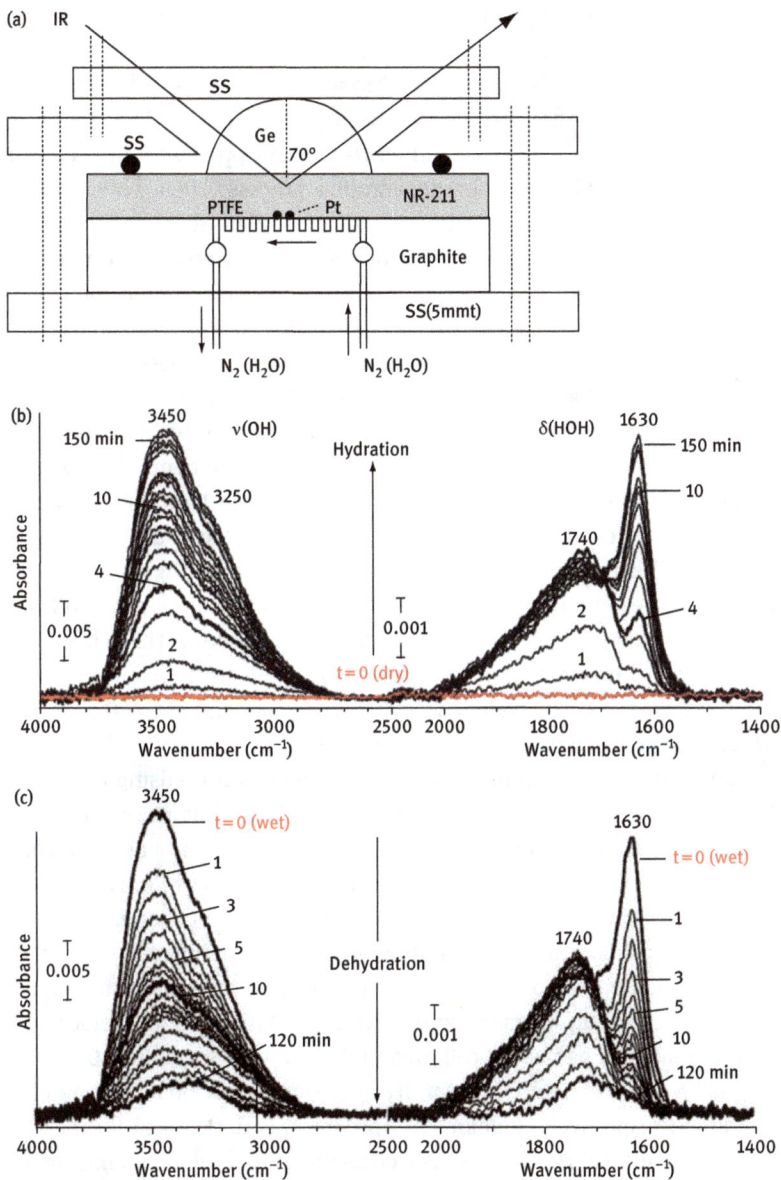

**Figure 5.4:** (a) Schematic showing the setup for ATR–FTIR and conductivity measurements of Nafion®
NR-211 during a hydration/dehydration cycle. Two platinum (Pt) wires (diameter 0.3 mm, length = 25
mm) probes were placed parallel with 1-mm spacing between a polytetrafluoroethylene (PTFE) mesh
and the membrane as for the conductivity measurement. A semi-cylindrical (20 × 25 mm) germanium
ATR prism, giving an angle of incidence of 70°, with an infrared (IR) penetration depth of 0.47 at 1,000
$cm^{-1}$ in the ATR measurements, was fixed on the membrane; and (b, c) change in $\nu$(OH) and $\delta$(HOH)
bands of water in Nafion® NR-211 during (b) hydration and (c) dehydration. Reproduced with
permission from K. Kunimatsu, B. Bae, K. Miyatake, H. Uchida and M. Watanabe, *The Journal of
Physical Chemistry B*, 2011, **115**, 15, 4315. ©2011, American Chemical Society [34].

was used to obtain background spectra. As a pre-treatment, the membrane was first dried by flowing dry $N_2$ (20 mL/min) through the gas flow field for 3 h and an ATR–FTIR spectrum was recorded. Then, the membrane was hydrated by supplying humidified $N_2$ gas at RT. After acquiring spectra at selected time intervals using rapid-scan mode for 2–3 h, the humidified $N_2$ was swapped with dry $N_2$, and measurements were continued during the dehydration process. Their work showed that hydration of a dry membrane initially results in complete dissociation of the sulfonic acid groups, resulting in the liberation of hydrated protons. These protons, secluded from each other, were identified as $\delta$(HOH) with a vibrational frequency $\approx$1,740 cm$^{-1}$. Their study also revealed two discrete components (1,740 and 1,630 cm$^{-1}$) comprising the $\delta$(HOH) band of water inside the membrane. These two components exhibited distinctive behaviours during hydration/ dehydration.

As seen in Figure 5.4b, a broad $\nu$(OH) band steadily evolved during hydration, whereas the $\delta$(HOH) band had two clearly separated components (1,740 and 1,630 cm$_{-1}$), each of which displayed different behaviour. The 1,740 cm$^{-1}$ component developed much faster as compared with that at 1,630 cm$^{-1}$, and reached a maximum at $\approx$4 min. This trend was reversed during dehydration (Figure 5.4c). The band at 1,740 cm$^{-1}$ represents the fully-hydrated state, but it can be seen shifting between $\approx$1,710 and $\approx$1,740 cm$^{-1}$ during hydration/dehydration. This shift was assigned by the authors, based on their previous work [35], to the change in the hydrogen bonding character of the water associated with the band at 1,740 cm$^{-1}$.

Hara and co-workers extensively investigated the water distribution inside Nafion® in an operating PEMFC and its dependence on temperature using a specialised cell to facilitate *in situ* micro-Raman spectroscopy of the interior of the membrane [36]. The cell consisted of an aperture of 500 μm on the cathode side fitted with a quartz window, and was used to allow laser-beam irradiation and Raman scattering. Heated air was blown onto the quartz window to prevent water condensation on it. The researchers used a 1,800-line mm$^{-1}$ grating for the confocal micro-Raman spectrometer fitted with a helium–neon laser (632.8 nm). The spot size was 2 μm (width) and 5 μm (depth). The cell setup utilised Nafion® NR-212 (60 μm) and 47.9 wt% platinum (Pt)/carbon (Tanaka), a geometric cell area of 3.61 cm$_2$, Pt loading of 0.5 ± 0.1 mg/cm$_2$, and a gas flow rate of 50 mL/min for dry and 20 mL/min and 40 mL/min for humidified $H_2$ and $O_2$, respectively. Raman spectroscopy was undertaken for the cell operating at 40, 60, 80 and 110 °C under $N_2$ as well as under fuel cell operating conditions with $H_2$ and $O_2$. The Raman peaks associated with the stretching vibrations of sulfonic acid groups (S-O), which were close to 1,058 cm$^{-1}$, and the change in its relative intensity with respect to symmetric stretching vibrations of C-F bonds (731 cm$^{-1}$) were meticulously studied. They observed that the hydration level of the membrane was higher in the operating cell (with $H_2$ and $O_2$) as opposed to when it was purged with $N_2$ despite using identical humidification levels. The shift in the S-O peak position and relative intensity along with the estimated values of $\lambda$ at different temperatures reported in this work by Hara and co-workers can be seen in Table 5.1. Their study also did not observe any shift in the peak positions of the other characteristic bands of Nafion® (C–O, S–C at 970 and 804 cm$^{-1}$, respectively).

**Table 5.1:** Variation in λ and S–O positions and relative intensities under different cell conditions.

| Cell condition | Cell temperature (°C) | S–O peak position (cm$^{-1}$) | $A_{(S-O)}$/ $A_{(C-F)}$ | λ (anode) | λ (cathode) |
|---|---|---|---|---|---|
| H$_2$/O$_2$ | RT | 1,058 | – | – | – |
| RH 30% | 60 | – | 0.28* | 3.4 | 5.2 |
| 20 mA/cm$^2$ | 80 | – | 0.255* | 3.6 | 4.9 |
| | 110 | – | 0.22* | 3.9 | 4.7 |
| Purged with N$_2$ | 40 | 1,057 | 0.3 | 2.9 | 4.9 |
| at RH 50% (λ | 60 | 1,056.4 | 0.275 | 3.6 | 5.1 |
| values | 80 | 1,056.1 | 0.25 | 3.9 | 5.2 |
| reported for RH | 110 | 1,057.5 | 0.2 | – | – |
| 70%) | | | | | |

Values estimated from graphs reported by Hara and co-workers
*at a depth of 30 μm from the anode
$A_{S-O}$/$A_{C-F}$: Relative intensities of S–O *versus* C–F
Adapted from M. Hara, J. Inukai, K. Miyatake, H. Uchida and M. Watanabe, *Electrochimica Acta*, 2011, **58**, 1, 449 [36]

In another recent study, Allen and co-workers undertook TEM and cryogenic TEM tomography on dry and hydrated 100 nm-cast Nafion®. They revealed spherical isolated ionic clusters (diameter ≈3.5 nm) corresponding to the hydrophilic sulfonic acid-containing phase in dry state. Cryo-TEM tomography enabled the first nanoscale three-dimensional views of the internal structure of hydrated Nafion® using direct imaging. The hydrated membrane consisted of an interconnected channel-like network with a domain spacing of ≈5 nm. Moreover, employing analytical TEM, the researchers could map the hydrophilic and hydrophobic domains in the dry membrane [37].

## 5.3 Combining *in situ* and physical characterisation methods: Membrane conductivity and the factors governing it

As we now know, membrane conductivity is dependent on proton mobility, which is governed by factors including hydration level and water channel geometry. The ATR–FTIR studies conducted by Kunimatsu and co-workers [34] explored hydration/dehydration in Nafion® (as discussed in the previous section) but they were combined with PC measurements. They recorded PC simultaneously with ATR measurements. The broad ν(OH) band was ascribed to several overlapping components (symmetric and asymmetric OH stretching bands, ν1(OH) and ν3(OH), and Fermi resonance of the δ(HOH) band with the ν(OH) fundamental) [38], so the researchers focused only on the δ(HOH) band and minimised ambiguity. Deconvolution of the δ(HOH) band resulted in three components (1,630, 1,740 and 1,840 $cm^{-1}$). The presence of a second component for the peak around 1,740 $cm^{-1}$ indicated that waters with different degrees of hydrogen bonding were associated with protons. The studies also found a linear relationship between the intensity of the band at 1,630 $cm^{-1}$ (which was ascribed to different degrees of hydrogen bonding between sulfonate groups and other water molecules) and the PC, as seen in Figure 5.5. On the other hand, the conductivity showed a negligible increase with the increase in the 170 $cm^{-1}$ band up to ≈0.4. Beyond this, the 1,740 $cm^{-1}$ component saturated while the conductivity increased rapidly with the 1,630 $cm^{-1}$ component. These two distinct behaviours in the conductivity trend were ascribed to i) distinct separation of the dissociation step of the $-SO_3^- -H^+$ groups, which would barely contribute to proton transfer and b) subsequent hydration of the membrane, which dominantly supports proton transfer. These findings were also in agreement with the density functional theory analysis by Paddison and Choe and co-workers [39, 40].

The influence of the geometry of water channels formed inside the Nafion® membrane and proton mobility on the conductivity of Nafion® has also been investigated [41]. Using impedance spectroscopy measurements coupled with a special cell for membrane and aqueous solutions, Schalenbach and co-workers compared the conductivities of fully-hydrated Nafion® membranes with those of aqueous hydrochloric acid with proton concentrations equal to those of fully-hydrated water channels in Nafion®. They found that the PC of aqueous HCl was approximately 6-fold higher than that of Nafion®. The decrease in water content naturally led to decrease in PC. The decrease in PC was attributed to the increased geometric restrictions faced by the protons due to higher tortuosity and reduced connectivity between the water channels at lower hydration levels. These researchers also suggested the possible hindrances to proton mobility caused by large-spacious anions constituting the functional groups and side chains, as well as inhomogeneous distribution of protons due to electrostatic attractions between the protons and anions at the walls of the water channels.

Apart from PC, attention must also be paid to thermal conductivity, which is known to decrease with increasing temperature (thermal conductivity for dry Nafion® at RT ≈0.16 W/m/K, decreases to ≈0.13 W/m/K at 65 °C). This phenomenon has been explained on the basis of phonon transport in crystalline polymers, wherein the possibility of collisions between the phonons increases with temperature due to the higher number of phonons involved, leading to a decrease in the mean free path. The presence of humidity at higher temperatures is, therefore, important because the thermal conductivity of water increases with temperature [42, 43], and it could assist in managing the decreasing thermal conductivity of the membrane.

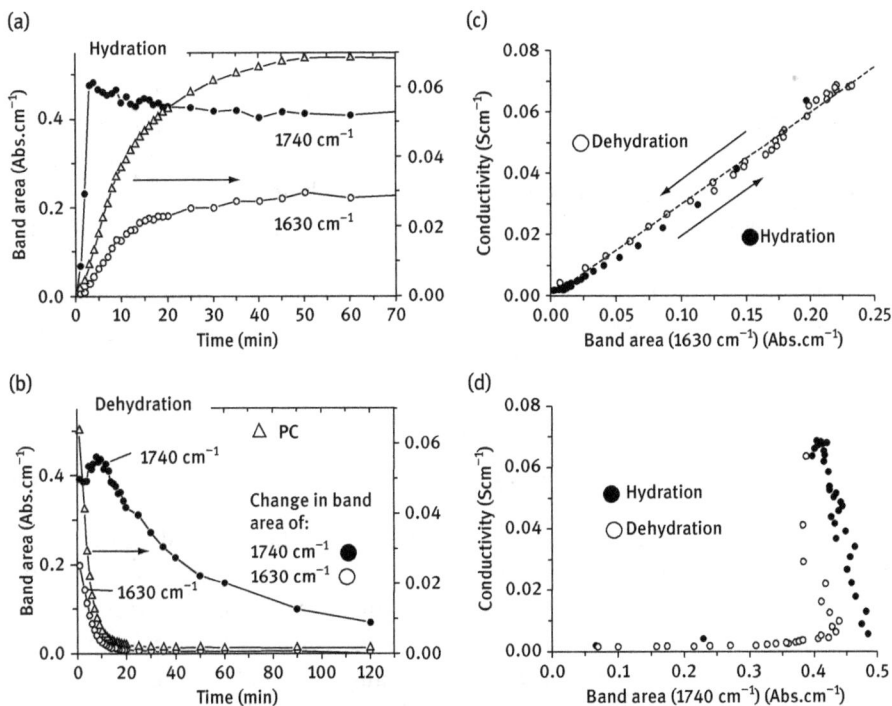

**Figure 5.5:** (a, b) Change of band area of the 1,740 cm$^{-1}$ and 1,630 cm$^{-1}$ components (as obtained from deconvolution) and PC of Nafion® NR-211 during (a) hydration and (b) dehydration; and (c, d) PC *versus* integrated area of the (c) 1,630 cm$^{-1}$ and (d) 1,740 cm$^{-1}$ components of the δ(HOH) band during hydration/ dehydration. Reproduced with permission from K. Kunimatsu, B. Bae, K. Miyatake, H. Uchida and M. Watanabe, *The Journal of Physical Chemistry B*, 2011, 115, 15, 4315. ©2011, American Chemical Society [34].

## 5.4 Combining *in situ* and physical characterisation methods: Gas permeation through Nafion® membranes

Gas crossover, which includes permeation of hydrogen from the anode to cathode and that of oxygen from the cathode towards the anode, is unavoidable in Nafion® and other PFSA membranes. Gas crossover occurs due to various processes, such as the intrinsic hydrogen/oxygen permeability of the polymer and through the water channels formed in the hydrated Nafion® membrane (inferred from the increase in hydrogen/oxygen permeability with increased water uptake) [44, 45]. Solubility of the gas in the medium (S$_{gas}$) together with the diffusion coefficient (D$_{gas}$) define the gas permeability (ε$_{gas}$) (Equation 5.1) [46]:

$$\varepsilon_{gas} = D_{gas}\,S_{gas} \tag{5.1}$$

For simplicity, the subscript 'gas' will not be used henceforth. Please note 'ε' was used to denote electro-osmotic drag earlier in Chapter 2.

Gas permeability in polymers is considered to occur through thermally-activated gas molecules jumping/permeating across the potential barriers formed due to the van der Waals forces between the polymer chains. The temperature influences the vibration energy of the molecules and contributes to the diffusion coefficient. The movement of the gas molecules is identified as random thermal movement or Brownian motion, which is responsible for diffusion.

The temperature (T) dependence of the gas diffusion coefficient (Equation 5.2) and the temperature dependence of gas solubility given by the Boltzmann distribution is shown in Equation 5.3 [47, 48]:

$$D(T) \approx D_0 (T)_e^{-E_A/(k_B T)} \tag{5.2}$$

$$S(T) \approx S_0 (T)_e^{-\Delta H_S/(k_B T)} \tag{5.3}$$

where '$E_A$' is the activation energy for the jumps across the potential barrier, '$k_B$' is the Boltzmann constant, '$D_0(T)$' is the proportionality factor which accounts for the distance between the equilibrium positions of this process and the vibration frequency of the molecule in the diffusion coordinate, '$H_S$' is the heat of solution, and '$S_0(T)$' is a proportional factor in small temperature ranges; both pre-factors $D_0(T)$ and $S_0(T)$ can have a negligible temperature dependence, which allows them to be approximated as constants.

Combining the three equations (Equations 5.1–5.3) shown above, the permeability of gases through the polymers can be approximated in Equation 5.4:

$$\varepsilon(T) \approx D_0 S_0 e^{-(E_A + \Delta H_S)/(k_B T)} = \varepsilon_0 e^{-E_\varepsilon/(k_B T)} \tag{5.4}$$

where $\varepsilon_0$ is the proportionality factor of the permeability and $E_\varepsilon$ its activation energy [47].

In a liquid (for gas permeating through water channels), the Brownian motion, drag, viscosity of liquid (water), and temperature dependence of the viscosity of the liquid together influence the permeation of gas molecules. The literature, however, seems to be divided when it comes to understanding the hydrogen and oxygen permeabilities in dry and fully-hydrated Nafion®. Schalenbach and co-workers carried out an extensive study to precisely measure the hydrogen and oxygen permeability of dry and fully-hydrated Nafion® and the influence of factors such as pressure, RH, and temperature on them. They carried out extensive systematic work using three electrochemical setups to carefully examine the effect of different conditions (Figure 5.6) [47]. The gases supplied to the cell flow fields were first humidified at 95 °C and then cooled to RT before entering the cell, which ensured that they were saturated with water vapour and additionally carried liquid water to prevent membrane dehydration. The authors also considered that the gases dissolved in water

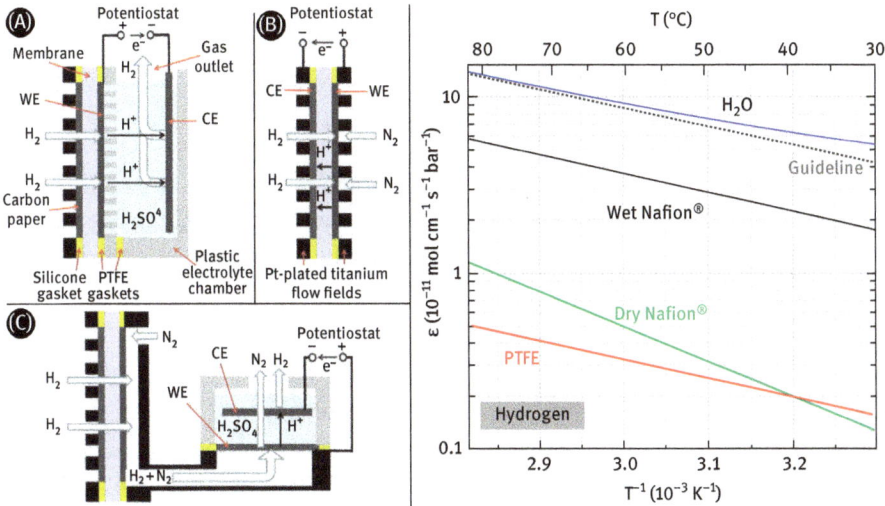

**Figure 5.6:** Schematic illustrations of the setups used for permeability measurements (A) Electrochemical half cells with aqueous sulfuric acid electrolytes and constant voltages were applied between the working electrode (WE) and counter electrode (CE) without using a reference electrode to enhance the measurement precision by reduced noise and capacitive currents. (B) Electrochemical cells in which the examined Nafion® membranes served as the electrolyte, similar to the setup of Sakai and co-workers [49]. (C) Setup used for dry Nafion® membranes in which the two cells were connected in series. The $H_2$ was purged (3 mL/min) in the first cell along one side of the membrane and $N_2$ (3 ml/min) on the other. The $H_2$ permeated through the membrane and was carried with the $N_2$ flux to the second cell. Reproduced with permission from M. Schalenbach, T. Hoefner, P. Paciok, M. Carmo, W. Lueke and D. Stolten, *The Journal of Physical Chemistry C*, 2015, **119**, 45, 25145. ©2015, American Chemical Society [47].

would lead to different viscosities and that, despite the two gases being non-polar and diatomic, the influence of their different molecular sizes on the gas–water system would be different. Consequently, they suggested that diffusion of two gases in the water channels must be considered in the form of mole fractions. Thus, the temperature dependence of the mole fraction ($\chi$) for hydrogen or oxygen dissolved in water at atmospheric partial pressure can be given as Equation 5.10 [46].

## 5.4.1 Gas permeation through water

Diffusion coefficient as a result of Brownian motion in solution under the influence of temperature is given as:

$$D = \mu k_B T \tag{5.5}$$

where 'μ' is the mobility of the molecule diffusing through the medium.

Considering the Stoke's drag, friction due to movement of gas molecules in the medium can be denoted as:

$$\frac{1}{\mu} = \beta C \eta \tag{5.6}$$

where 'η' is the viscosity of the medium, 'β' accounts for the geometry of the molecule, and 'C' is the Cunningham correction factor describing continuum effects.

The Andrade relation (Equation 5.7) denotes the temperature dependence of viscosity of water [50]:

$$\eta(T) = \eta_0 T e^{E_\eta/(k_B T)} \tag{5.7}$$

where '$E_\eta$' denotes the corresponding activation energy and '$\eta_0$' an empirical constant.

Combining the first two equations to derive the Stokes–Einstein equation and incorporating the Andrade equation into it we obtain:

$$D(T) = \frac{k_B}{\beta C \eta_0} e^{-E_\eta/k_B T} = D_0 e^{-E_A/(k_B T)} \tag{5.8}$$

When determining solubility, the mole fraction of the gas dissolved inside the water should be considered:

$$\chi \approx \frac{C_{gas}}{C_{H_2O}} \tag{5.9}$$

where 'c' is the concentration and for '$c_{gas}$ >> than '$c_{H2O}$' the above equation is true.

The temperature dependence of the mole fraction of oxygen and hydrogen dissolved in water at $p_0 \approx 1$ bar is given as [46]:

$$\chi(T) = \exp\left\{-S_1 + \frac{S_2}{T} + S_3 \ln[T/(100\,K)]\right\} \tag{5.10}$$

the values of $S_1$, $S_2$ and $S_3$ have been reported in the literature [46].

Thus, the temperature dependence of gas permeability of water can be denoted as [47]:

$$\varepsilon(T) = S(T)D(T) \approx \varepsilon_0(T)\varepsilon^{-(E_\varepsilon)/(K_B T)} \tag{5.11}$$

Using the three setups and the gathered data on the function of partial pressure, RH and temperature in defining gas permeation, Schalenbach and co-workers reported two main points, as shown below:

a) Role of partial pressure – setup B arranged with fully-hydrated Nafion® 117 coated both sides with catalyst (Pt/carbon 1 mg/cm²) was used to study the effect of pressure on hydrogen permeation flux at 80 °C. Fully-humidified $H_2$

(3 ml/min) was purged along the CE while fully-humidified $N_2$ (3 ml/min) was supplied to the other side. The partial pressure $pH_2$ was determined by deducting the calculated $pH_2O$ with reference to the cell temperature from the absolute pressure 'p' measured. The absolute pressure on the $N_2$ side was defined as the 'counter pressure'. The data collected in terms of $H_2$ permeation flux density, $\Phi_{H2}$ through Nafion® (determined from current density 'j' using Faraday's law where, $\Phi_{gas} = j/zF$ and 'z' is the number of electrons involved in the electrochemical conversion of the gas) as a function of $p_{H2}$ (for $1 \leq p_{counter} \leq 5$) revealed a constant slope, implying that permeability was independent of the counter pressure. At balanced and differential pressure, the $\Phi_{H2}$ through Nafion® depended only on the applied hydrogen pressure. The applied differential pressures, when ≤100 bar (i.e., smaller than the capillary pressures in the water channels [45]), did not drive the hydrogen through the water channels, suggesting that the water channels of Nafion® do not act as a porous media (as described by Darcy's Law) in this pressure range. Thus, the hydrogen permeation under these conditions was diffusive, whereby dissolved gases permeate through the Nafion® membrane *via* Brownian motion.

b)  Role of RH – to study the effect of RH, Nafion® NR-212 was used in setup A and B at 80 °C. Due to the limitation of the conductivity of the membrane, the setup was not used for a dry membrane. Measurements were carried out at the atmospheric pressure of the supplied gases (purged 50 ml/min through the flow fields). The temperature of the humidifers was increased at 0.1 K/min to achieve a range of RH. The authors reported that the hydrogen permeability increased as the RH increased. Moreover, saturated water vapour with dissolved hydrogen at RH = 100% resulted in ≈3.5% less permeability than that for the gas mixture which additionally carried liquid water. This behaviour may be ascribed to the different types of water content inside Nafion® at these conditions, as explained by Kreuer [51]. Further simulation studies using resistor networks suggested an increase in the permeability of the solid phase in Nafion®, which could be attributed to a softening of the polymeric matrix resulting from water uptake. It was suggested that water may be acting as a plasticiser for the polymeric matrix, similar to the softening of the viscoelastic properties of Nafion® reported by Kreuer [51]. To explain this further, the swelling of Nafion® (as a consequence of water uptake) increases the mean distance 'r' between the polymer chains. This may reduce the van der Waals forces (proportional to 1/r Lennard–Jones potential, as discussed by Allen and co-workers [37]) between the polymer chains in the solid phase. Further influx of water into the plasticised polymer network could weaken the van der Waals forces between the polymer chains inside the solid phase. The inclusion of softening of the solid phase enabled better agreement of the simulated models with the experimental data, and led to the deduction that the softening effect may dominate the increase in the overall permeability. This work also suggested the possibility of an intermediate phase during which the van der

Waals forces between the aqueous and solid phase are weaker due to a less distinct H-bond network, which results in larger diffusion coefficients and solubilities for $H_2$ and $O_2$. Considering the three-phase (aqueous, solid and intermediate) mixed pathways for the movement of gas molecules, the overall permeability of Nafion® was estimated to increase by a factor of 5.4 in comparison with the simulated permeability of the aqueous phase alone. This led to an estimated ≈81% of the $H_2$ permeability in fully-hydrated Nafion® resulting from permeation through the intermediate phase and solid phase [47].

These studies implied that gas molecules can take 'shortcuts' through the solid phase when permeating through the Nafion® membrane but protons cannot (Figure 5.7). Consequently, the diversions for proton transport through Nafion® are larger than those for gas molecules.

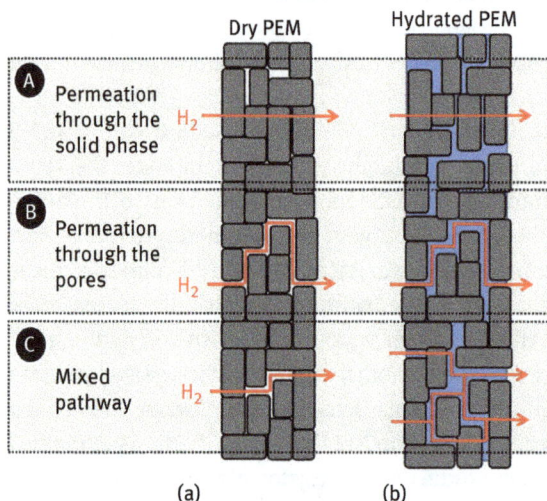

**Figure 5.7:** Graphic sketch of the pathways (arrows) for gas permeation through a section of a PEM for hydrogen molecules. The grey area depicts the solid polymeric phase (agglomerate of ionomers), blue represents water, and white represents gas- filled pores. a) Dry PEM and b) hydrated PEM. Permeation of gas molecules through, (A) the solid phase, (B) membrane pores or water channels and (C) mixed pathways as a combination of alternating permeation through the aqueous and solid phases of PEM. Reproduced with permission from M. Schalenbach,T. Hoefner, P. Paciok, M. Carmo, W. Lueke and D. Stolten,*The Journal of Physical Chemistry C*, 2015, **119**, 45, 25145.©2015, American Chemical Society [47].

c) Role of temperature – setup A and C were utilised to study the influence of temperature on $H_2$ and $O_2$ permeability through Nafion® membranes of various thicknesses (Nafion® NR-212, 1135, 115 and 117), where fully-hydrated membranes were subjected to different absolute pressures of fully-humidified

$H_2$ (2, 3, 4 and 5 bar). The temperature dependence of gas permeability in an Arrhenuis plot appeared approximately linear. The difference in the permeabilities of $H_2$ and $O_2$ were attributed to the difference in the size and weight of the two molecules.

d) Apart from these factors, the different solubility of the two gases in hydrated and dry membranes also has a role. This factor, however, is not completely understood due to insufficient data in the literature on the solubility and diffusion coefficient of hydrogen in PTFE. It has been reported that in fully-hydrated Nafion®, the trend for permeability is similar with $H_2$ showing 18% greater activation energy than that of $O_2$. In the case of a dry membrane, the approximate activation energy for $H_2$ permeability is smaller than that of $O_2$ [47].

Based on these results, there are three approaches for membrane and polymer designing recommended for reducing gas permeation while maintaining good PC. The first approach involves having higher concentrations of functional groups inside the water channels to minimise the solubility and diffusion coefficients of the gases in the aqueous phase. The second approach is to increase the intermolecular forces between the aqueous and solid phase to reduce permeation through the intermediate phase. The final approach is to use crosslinked ionomers to prevent the solid phase softening and the avoidance of dead-end water channels inside the membrane because they do not contribute to proton conduction and may further assist gas permeation through mixed pathways [47].

## 5.5 Membrane inside the operating proton-exchange membrane fuel cells: influence of catalyst layer defects

Various reports have pointed out that the mechanical properties of pure PFSA membranes and CCM can differ widely. The restraints imposed upon the membrane due to the catalyst and diffusion layers are shown to inhibit in-plane expansion and contraction, resulting in decreased tensile peak stress as compared with the pure membrane and, during hydration, the through-plane stress and strain increase [52]. The tensile and dynamic mechanical properties of CCM have also been reported to be more sensitive to temperature than RH variations. Consequently, CCM can suffer from various morphological defects, such as catalyst layer cracks, delamination, electrolyte clusters, orientation, thickness variations, and Pt clusters [53].

Tavassoli and co-workers carried out a systematic study on customised MEA with artificially designed and fabricated defects to determine possible correlations between pre-existing catalyst layer defects (primarily, cracks, contamination and delamination) and local membrane degradation. They used a 5-cell stack with 4 customised MEA. The cells were separated by graphitic bipolar plates and the stack was evaluated using the cyclic open circuit voltage (COCV) AST protocol defined by Ballard Power Systems [54] to expose the membranes to *in situ* chemical and mechanical stresses specific for automotive applications. The three types of defects were introduced into the CCM in three ways, as detailed below:

- Cracks – the catalyst layer was scrapped off from the anode or cathode gas diffusion electrode (GDE) surface at regular intervals in circular areas (diameter 1–2.5 mm) to simulate cracks.
- Contamination – Iron (Fe) contaminations (100, 200 and 1,000 ppm mg Fe/kg of membrane) were introduced into the membrane at periodic positions using an aqueous solution of $Fe_2O_3$. The membrane was dried at ambient temperature before being hot pressed into an MEA.
- Delamination – two types of delaminations were created: delamination with a bare membrane, and delamination with residual catalyst particles on the membrane surface. A porous (pore size: 0.8 μm), thick (9 μm), hydrophilic polycarbonate film was inserted between the membrane and the catalyst layer to simulate delamination. A small amount of catalyst ink was sprayed on one side of the polycarbonate film to simulate delamination with residual catalyst particles.

Monitoring of OCV changes to correlate them with $H_2$ crossover and *post mortem* scanning electron microscopy (SEM) was carried out on the MEA (for *post mortem* SEM, MEA were cut and embedded into epoxy) to study the effects of the defects introduced. The authors reported that cells with Fe-contaminated MEA failed earlier than those without contaminants. Moreover, the variable contaminant concentrations initially deposited were no longer traceable in the end-of-test MEA in which the membranes were globally thinned membrane with randomly distributed failure points. These confirmed accelerated chemical degradations in the membrane due to Fe ions acting as local initiators. In simulated delamination studies, the authors found a minimal effect of cathode delamination on membrane degradation whereas anode delamination led to severe local thinning. This anodic degradation was attributed to possible facilitated rapid release of $H_2O_2$ due to a hydrophilic anode environment and transfer into the membrane [55].

In another defect simulation study, Jung and co-workers artificially created pinholes of different sizes (diameter 400–540 μm) and at various locations (side inlet, outlet and middle of the anode) using a micro-needle (outer diameter 470 μm), thereby simulating pre-existing pinholes, to study their effect on $H_2$ crossover during PEMFC operation. They did single-cell studies using Gore PRIMERA® 57 MEA with an active area of 25 cm². The cell operating conditions were 65 °C, 100% RH (anode and cathode), and $H_2$ and air stoichiometries set at 1.5 and 2.0, respectively. Current density was 0–1.0 A/cm². As expected, the hydrogen crossover rate of the punctured MEA under all current densities was higher than that in a normal MEA. Moreover, the crossover rate in pricked MEA at OCV condition was considerably increased from the average $H_2$ concentration at other current densities because the gas could be transported directly through the pinhole in the absence of water in the pinhole. In terms of the location of the pinhole, it was reported that pinholes near the inlet and outlet at the anode resulted in lower performance and higher crossover rate, suggesting that the faults near the gas inlet and outlet influenced the performance and durability more with respect to middle locations due to the increased and greatly scattered $H_2$ crossover concentration. It is, therefore, recommended that membranes should be manufactured such that their inlet and outlet parts are stronger [56].

Alavijeh and co-workers investigated the influence of *in situ* hygrothermal fatigue on the microstructure and mechanical properties of Nafion® 211 (non-reinforced)-based CCM. They custom developed two AMST involving rapid cycling between the wet and dry states of the membrane to carry out systematic exploration of the fatigue process in a 5-cell stack. Partially degraded CCM were periodically extracted after each 4,000 cycle for *ex situ* characterisation. The end-of-test CCM (after 20,000 cycles) was further studied using *ex situ* methods. The cells had an active area of 45 cm$^2$. Furthermore, to completely eliminate the effects of chemical exposure and degradation, $N_2$ gas was used and any electrochemical diagnostics during the *in situ* experiments were avoided. To develop an understanding of the role of pure mechanical fatigue stress on the *in situ* PFSA degradation, the tensile and time-dependent hygrothermal properties recorded were compared with the previously reported mechanical properties of CCM, which were subjected to combined chemical and mechanical degradation [57]. The AMST-1 (AMST-2) were carried out for 20,000 cycles or until crossover was >10 sccm – whichever happens first; at 80 °C (95 °C) operating temperature, $N_2$ supplied on both sides at 9 slpm (3.5 slpm), ambient pressure. The wet and dry cycles of AMST-1 consisted of 2 min dry state (≈0% RH) + 2 min wet state (≈90% RH), whereas for AMST-2 they consisted of 3 min dry state (≈0% RH) + 1 min wet state (≈100% RH). Different duration of wet and dry cycles were chosen in the two AMST as the wet conditions cause compressive stress while the dry conditions at low RH lead to high tensile stress (maximised *via* AMST-2) due to the constrained contraction of the membranes inside the cells [58].

IR imaging of the end-of-test MEA for AMST-1 and AMST-2 identified membrane degradation and transfer leaks at the inlet region because this region inside the cells was exposed to slightly higher gas temperatures as compared with the other (middle and outlet) regions, and these regions were further examined using SEM (Figure 5.8). Membranes degraded *via* exposure to combined chemical and mechanical degradation using a separate COCV were compared with end-of-test CCM after the AMST. Cross-sectional SEM studies of the membranes (MEA embedded in epoxy and polished using 120–200 silicon carbide grit paper) revealed thinning of ≈4–5% and ≈9%, in AMST-1 and AMST-2, respectively. However, an extensive 48% thinning was observed for membranes exposed to the COCV AST, confirming that combined mechanical and chemical stress can drastically increase degradation. The degradation in AMST was attributed to in-plane viscoplastic deformation and residual stresses bearing in mind the dominant fatigue

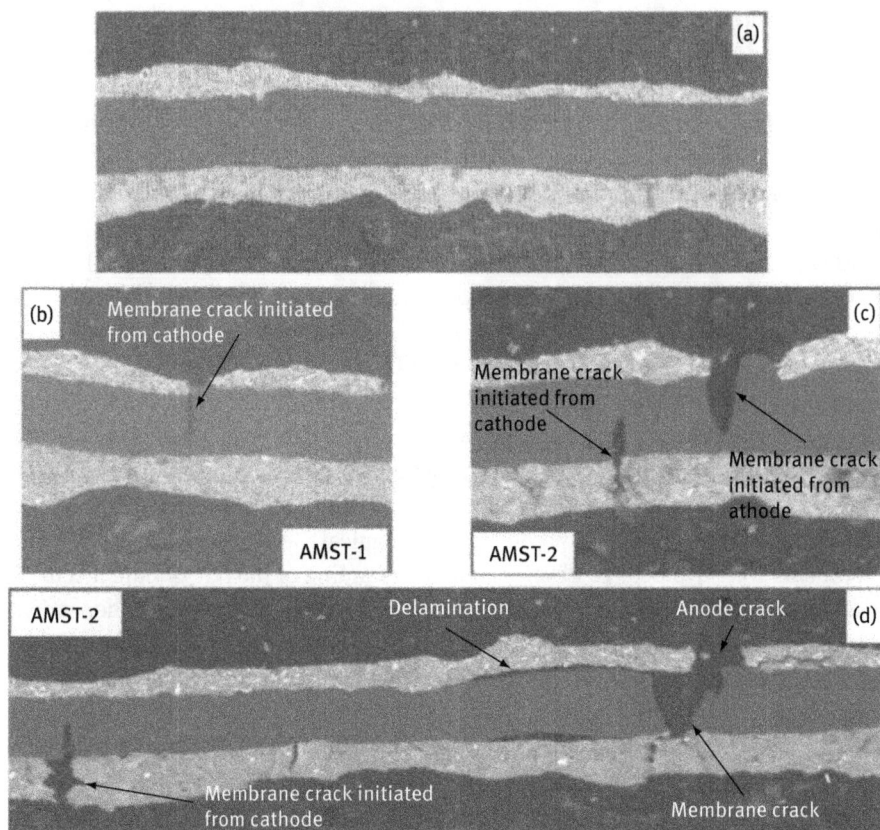

**Figure 5.8:** Cross-sectional SEM images of MEA at the (a) beginning-of-life (BOL) and (b–d) end-of-test from (b) AMST-1 and (c, d) AMST-2. Reproduced with permission from A.S. Alavijeh R.M.H. Khorasany, Z. Nunn, A. Habisch, M. Lauritzen, E. Rogers, G.G. Wang and E. Kjeang, *Journal of Electrochemical Society*, 2015, **162**, 14, F1461.©2015, Electrochemical Society [57].

stress in this direction [59]. The higher stress amplitude and temperature during AMST also led to more significant thickness reduction during this test. Mechanical damage was observed in the form of cracks and delamination. Major cracks appeared to extend from the membrane into the adjoining catalyst layers, indicating substantial interaction between these two components. The membrane leaks and failures identified *via* IR imaging could be attributed to occasional, large cracks spanning the entire membrane thickness (Figure 5.9). Formation of small microcracks inside the membrane and delamination, possibly resulting from material fatigue, can also be seen in Figure 5.9. Naturally, more frequent and relatively larger microstructural damage was observed for membrane subjected to AMST-2. *Ex situ* tensile testing carried out on CCM (using a standard 5:1 length:width aspect ratio specimen) removed from the AMST at different stages revealed that the

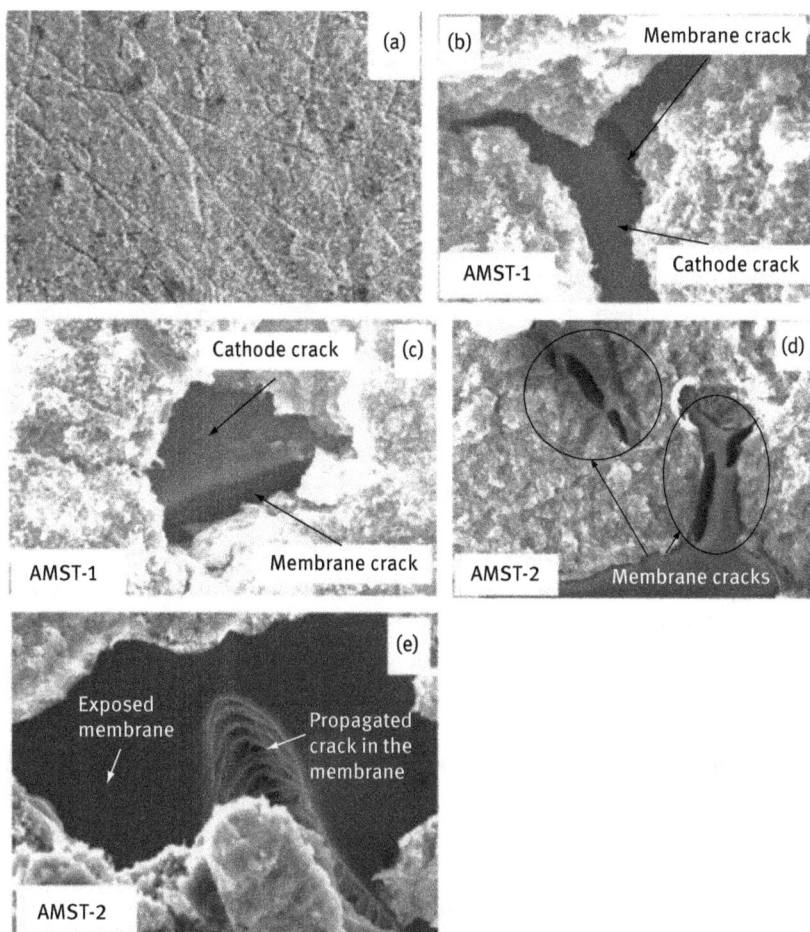

**Figure 5.9:** Top surface SEM images of CCM at (a) BOL and (b–e) end-of-test from (b, c) AMST-1 and (d, e) AMST-2. The samples were taken from the severely damaged inlet areas revealed by IR imaging. Reproduced with permission from A.S. Alavijeh R.M.H. Khorasany, Z. Nunn, A. Habisch, M. Lauritzen, E. Rogers,G.G. Wang and E. Kjeang,*Journal of Electrochemical Society*, 2015, **162**, 14, F1461.©2015, Electrochemical Society [57].

final strain (after maximum elongation of 26 mm, i.e., 330% from an initial length of 6 mm because the membranes did not reach fracture point) was reduced appreciably (≈300% at 4,000 cycles to 20–55% at 20,000 cycles). The rate of decay in elongation was reported to be relatively high for up to 12,000 cycles, after which it slowed down until 20,000 cycles. These findings suggested that the microcracks initiated inside the membrane in the early stages of the AMST had greater influence on its ductility than the propagation of cracks during more mature stages of degradation [57].

## 5.6 Membranes in exceptional conditions (durability and degradation): Challenges under application and operation-specific conditions

Durability of PEMFC needs to be established under a wide range of operational conditions for them to be commercially viable for automotive as well as other stationary and portable applications. Emphasis on such studies is slowly gaining momentum. Each type of application presents itself with unique climatic, environmental and operational conditions. Obviously, the suitability and adaptability of proton-conduction membrane under these exclusive conditions forms an important part of such durability studies.

### 5.6.1 Membranes in cold start-up systems and freeze/ thaw cycles

Cold start or start-up from sub-zero temperatures is an essential requirement for commercial application of PEMFC, especially in transportation-related applications. The United States (US) Department of Energy (DOE) technical target for 2020 includes the ability of the fuel cell stack to reach ≤50% of its power within 30 s at −20 °C [60]. Water inside the fuel cell system has a defining role in such freezing conditions. Inside a PEMFC, water can exist in various phases and go through phase transitions during different operating temperatures when subjected to freezing conditions. The state of water also depends on the media/component of PEMFC in question (Table 5.2). Under freezing conditions, the volume of water expands by ≈9% as the water transforms from ice to liquid due to the difference in the densities of water (density = 999.8 kg/m$^3$) and ice (916 kg/m$^3$) in these two phases. Naturally, the freezing of liquid water will result in the generation of unbalanced stresses. Repetitive cycles of generation and release of stress (as the ice thaws and freezes) can significantly impact the membrane structure and result in structural damage [61]. F/T and cold-start studies done in combination with *in situ* neutron tomography, differential scanning calorimetry (DSC), X-ray and NMR studies over the last few years have revealed interesting results suggesting that water inside a membrane can exist as vapour, liquid water and ice [62–69]. There is a lack of consensus on the aggregate states in which water exists inside a membrane, but there is consensus that, at the beginning of a cold start, the membrane takes up a certain amount of water, which is identified by a drop in ohmic resistance. After a certain time, the membrane resistance stabilises, which is explained by saturation of water content inside the membrane [70]. Other studies using DSC and NMR have also indicated that water does not freeze until λ > 4.8. Studies of bond strengths between water and sulfonic groups have classified water contents

**Table 5.2:** Phases of water present in different components of PEMFC during cold-start processes.

| Component | Location | State of water |
|---|---|---|
| PEM | Ionomer | Ice, liquid water, non-frozen water, frozen water and vapour |
| GDL | Pores | Ice, liquid water and vapour |
| Cloisite® | Pores and ionomer | Ice, liquid water, non-frozen water, frozen water and vapour |
| Flow channel | Everywhere | Vapour and liquid water |

GDL: Gas diffusion layer
Reproduced with permission from Z. Wan, H. Chang, S. Shu, Y. Wang and H. Tang, *Energies*, 2014, **7**, 5, 3179. ©2014, Z. Wan, H. Chang, S. Shu, Y. Wang and H. Tang and MPDI [61]

as 'non-freezable' ($\lambda \leq 4.8$), 'freezable' (water molecules loosely bound to sulfonic acid groups) and 'free water' (i.e., when water content is relatively high, also called 'surface water') [67, 71, 72].

McDonald and co-workers undertook extensive *ex situ* F/T cycling (−40 to +80 °C) over 3 months (385 cycles) on Nafion® 112 (and MEA) and observed key changes, including a decrease in anisotropy of tensile strength and water swelling behaviour and lower oxygen permeability along with improved through-plane conductivity in constrained MEA (constrained along the X, Y and Z directions to simulate real fuel cell stack arrangement) under low RH values. They concluded that a molecular-level rearrangement involving the manner in which the sulfonic acid groups are arranged takes place as a result of F/T cycling. Cycling between −20 °C and RT also revealed several cracks due to unbalanced stresses [73].

Interestingly, in a simple approach, Cho and co-workers demonstrated that purging the PEMFC cell with dry gas (to remove water) or filling the compartments with anti-freeze solution before the cell temperature fell to 0 °C can prevent MEA degradation [74].

Limited information is available on the effect on the membrane during F/T cycling in DMFC. DMFC have more water in their system because the methanol is supplied as an aqueous dilution. Consequently, there is plenty of water in the anode as well. Here, membrane thickness also has a role in degradation during F/T cycling. A study on DMFC was done using Nafion® 112, 115 and 117 for F/T cycling between −32 and +60 °C by Oh and co-workers [75]. They reported that DMFC cells with thinner membranes suffered from damaged triple-phase boundary and loss of electrochemically-active surface area (ECSA) due to F/T cycling. This was attributed to reduced cathode catalytic activity as a consequence of methanol crossover, which was aggravated in the case of thinner membranes after continuous F/T cycling. Cells with thicker membranes (Nafion® 115 and 117) also showed a more restrained loss of power density (36 and 28%, respectively) as opposed the thin Nafion® 112 membrane (69%) as observed after 20 F/T cycles. Methanol crossover resulting in membrane fracturing and eventual cathode poisoning is one of the major challenges under extreme temperature conditions for DMFC.

## 5.6.2 Open circuit conditions

Open circuit operation studies define the durability and reliability of PEMFC under idle conditions. This is a highly practical element for the commercial feasibility of this power generation system, especially for automotive/transport and portable applications, where the PEMFC system may remain idle for several thousand hours during its lifetime. Near open circuit conditions in which the system is subjected to only an idle auxiliary-load power withdrawal at a relatively high-PEMFC potential ($\approx$0.9 V) are also an equally impelling stressor. Wu and co-workers did an AST for 1,200 h under close to open circuit conditions using a small 6-cell PEMFC stack and analysed the cells and MEA using cyclic voltammetry, EIS, linear sweep voltammetry (LSV) and direct gas mass spectrometry (DGMS). The cells, with active area of 50 cm$^2$, consisted of Gore PRIMERA® 57 CCM with 0.4 mg$_{Pt}$/cm$^2$ of Pt/carbon loading. The stack operated at a constant low current of 0.5 A (10 mA/cm$^2$) with a consequent cell voltage of $\approx$0.9 V (0.914 V) at the BOL. The stack was operated at 70 °C with fully-humidified air (2.0 slpm) and hydrogen (0.5 slpm) at stoichiometries of 2.5 and 1.5, respectively. The authors compared the polarisation curves for the 6 cells individually at the BOL and reported <1% difference in performances compared in the range of low current densities and only 0.67% when stack current density was controlled at 10 mA/cm$^2$. As the stack testing began, they observed noticeable changes in performance after the first 800 h of operation. The degradation rate for the stack (in terms of voltage decay) for the first 800 h of operation was $\approx$0.119 mV/h (decreased from 0.914 to 0.819 V). After $\approx$800 h, the rate of decay more than doubled, reaching 0.260 mV/h (from 0.864 to 0.760 V in 400 h). A significant decrease was also observed in the ECSA after 800 h (from 0.037 to 0.021 C/cm$^2$). The loss of the polymer resin could possibly lead to the unavailability of some Pt catalyst for the electrochemical reaction and consequent loss of ECSA of the catalyst layer. This was further clarified with more speedy decay of ECSA (from 0.021 to 0.002 C/cm$^2$) along with a drastic increase in the H$_2$ crossover rate (increased from 1.84 mA/cm$^2$ at BOL to 2.15 mA/cm$^2$ at 800 h and; 9.54 mA/cm$^2$ at 1,000 h to 20.71 mA/cm$^2$ at 1,200 h), which was measured using hydrogen oxidation current density (obtained between 300–350 mV) from the LSV measurements. The DGMS analysis carried out using cathode exhaust gas further supported the evidence of degradation. The DGMS signals detected molecular weights, which were assigned to hydrogen fluoride, carbon monoxide, N$_2$, hydrogen peroxide and carbon dioxide (at 539 h) and an additional signal for sulfurous acid at 1,052 h. Hence, while the initial loss of potential was due to small amounts of hydrogen crossover, as the rate of crossover increased with the passing operation hours, unwanted chemical reactions at the anode and cathode (along with the formation of hydrogen peroxide) led to generation of the radical species $HOO^{\bullet}$ and $HO^{\bullet}$. These fed into the destructive cycle/loop, accelerating chemical degradation of the membrane, resulting in membrane thinning and more severe gas crossover. With more intense gas crossover, exothermic direct reactions between $H_2$ and $O_2$ began on the catalyst surface, leading to hotspots (identified by release of sulfurous acid) and local pinholes, resulting in further thermal degradation of the membrane. This eventually led to failure of the stack. Thus, the combined analysis of the results reported by Wu and co-workers revealed that the initial loss of stack performance was mainly due to catalyst decay but the gas crossover and accelerated formation of radical species during OCV operation led to the subsequent intense degradation as a result of membrane failure.

Another OCV study undertaken using a bilayer (twin CCM) membrane system in a 4-cell stack for 1,600 h was reported by Zhang and co-workers. They aimed to facilitate *post mortem* analysis of the membrane after degradation (detachable membrane structure). An additional benefit of this configuration was an identifiable initial position of the membrane centre. They used pairs of four membranes (Nafion® 117, 115, 212 and 211) in the four cells forming the stack. Each membrane was a CCM (Pt loading 0.3 mg/cm² in the catalyst layer) coated only on one side with the uncoated sides facing each other. The MEA were formed by sandwiching the twin-CCM (sealed using Kapton® film) between the anode and cathode GDL. They also tested the cell under similar conditions (70 °C with fully-hydrated gases using identical stoichiometries; active area 50 cm²), just like Wu and co-workers. The overall performance of the stack was relatively low, which was most likely due to the higher resistance of the double-layered membranes. Not unexpectedly, their report identified disproportionately high OCV loss and hydrogen crossover for the two thinnest membranes (rapid OCV decrease for cell 4, which housed the thinnest membrane, after ≈1,400 h, followed by cell 3, which had the second thinnest membrane, after 1,600 h due to membrane thinning, hotspots and pinhole formation through thinner membranes), which was further confirmed by *post mortem* IR measurements and SEM. Cross-sectional SEM indicated more thinning at the cathode side. Atomic force microscopy images revealed significantly stiffer surfaces for the aged inner membrane, especially for the thinner membranes, and had a severely degraded surface structure with polymer flakes [76].

### 5.6.3 Effects of vibrations

Vibrations are one of the common sources of mechanical damage in transportation systems. Based on existing transportation vehicles, the different types of vibrations that the PEMFC system may experience are vibrations induced due to: unevenness of the road surface; oscillations of the axel and wheel with the suspension system (8–16 Hz); auxiliary devices in semi-trailers to feed systems such as fans and compressors (0.9–5.8 Hz) due to road conditions; auxiliary devices subjected to vibrations (17–40 Hz) due to internal combustion engine vibrations [77–80]. The maximum amplitude of vibrations in transportation applications generally is ≤0.95 g, where g is the acceleration due to gravity (9.81 m/s$^2$) [76]. Earlier reports exploring short-duration studies on PEMFC systems, including subjecting the stack to horizontal and vertical vibrations (30–150 Hz for 90 mins) and the effects of underground mining conditions (shock and vibrations), did not reveal damage to membranes leading to cracks [81–84]. Recently, Hou and co-workers undertook vibrational studies using a 9-kW fuel cell stack (assembled with 90 cells of 250 cm$^2$ effective area in series) to investigate the variations of performance under strengthened road vibration. They mainly investigated the overall degradation of the stack, but they also reported that the stack OCV experienced slight decay during the vibration tests, and concluded that the current losses were due to $H_2$ crossover and short-circuit current increase under long-term vibration conditions [85].

Banan and co-workers carried out simulation studies to investigate the effect of mechanical vibrations on delamination propagation at the membrane–Cloisite® interface. They simulated a PEMFC with 25 μm-thick Nafion® membrane consisting of an initial delamination length of 'a' = 25 μm on the cathode side. The temperature and humidity conditions were set at 85 °C and 90% RH, respectively, to simulate the operating conditions of a fuel cell. Crack propagation in the fuel cell (subjected to AST for 300 h) was examined for different vibration frequencies ($\omega$ = 5, 10, 20 and 40 Hz) and amplitudes (A = 1, 2, 3 and 4 g). Banan and co-workers revealed five interesting results, which are summarised below:

1. The amplitude and frequency of the vibration had a significant impact on delamination propagation.
2. Increasing the amplitude of the vibration had a non-linear effect on the delamination length. However, at larger amplitudes, this effect faded and was dominated by the effect of increasing frequency, which had a more severe impact on final delamination length. For example, at 10 Hz, increasing the amplitude from 2 g to 3 g led to an increased final delamination length from 0.51 to 0.74 mm, whereas the final delamination length only increased from 0.74 to 0.81 mm when increasing the amplitude from 3 to 4 g.
3. The simulation case with A = 4 g and $\omega$ = 40 Hz resulted in the most severe damage. No damage was observed for cases A <0.5 g because the applied vibrational force was insufficient for crack propagation within the 300-h simulation time.
4. Increasing the initial delamination length, 'a', significantly increased the rate of propagation. For A = 1 g and $\omega$ = 5 Hz, increasing 'a' from 0.025 to 0.250 mm increased the final delamination length from 0.13 to 1.05 mm. This behaviour was more significant for higher amplitudes and frequencies (for A = 4 g and $\omega$ = 40 Hz, where delamination length increased from 1.46 to 2.86 mm).
5. Aggressive vibration conditions can dominate the effect of the initial delamination length on the final damage. For example, a final delamination length

for A = 4 g and ω = 40 Hz and a =0.025 mm was higher than that for A = 1 g and ω = 5 Hz with a larger initial delamination length (a = 0.250 mm) [86].

### 5.6.4 Aircraft applications: Climatic challenges

If PEMFC are to be considered for aircraft applications, the operational require-ments will have to be much more stringent as compared with road/land-based transportation, especially in terms of performing long-term at sub-zero tempera-tures apart from cold start-up. The applications for airborne transportation systems also require significantly larger fuel cell stacks. Stacks also suffer from inhomoge-neity of temperature distributions between the cells at the centre and those at the end of the stack. Moreover, aircraft-based applications also require thorough study of the effect on the various ancillaries and peripheral equipment dedicated to fluid and energy conditioning which can impact the operation of the fuel cell under such extreme conditions.

Bégot and co-workers did specially designed tests (in a climatic chamber) in accor-dance with the requirements and applications in an airplane. The tests were carried out on a 52-cell stack using commercial membranes (type not specified) with an active cell area of 156 cm$^2$. Among the different tests, the low-temperature tests are of particu-lar interest with respect to the membrane. At the beginning of low-temperature test, the surrounding temperature (inside the climatic chamber) was close to +15 °C whereas the stack and gas inlet temperatures were set at +60 °C. The gases were humidified (RH = 90%) and the stack was operated at a specific load (25, 50 or 75 A). The tempera-ture of the climatic chamber was then slowly reduced to −34 °C in steps of −5 °C (the temperature of the inlet streams of humidified gases remained close to +60 °C, no other stack heating was used). Once the desired temperature was reached it was en-sured that the stack power output was stabilised before polarisation curves could be recorded. The authors found that the thermal power delivered by the stack was suffi-cient to heat the fluid in the temperature circuit to the requested level of 60 °C. Never-theless, it was quite difficult to maintain gas temperatures of 65 °C at the gas inlets of the fuel cell stack, and the hydrogen and air inlet stream temperatures were actually close to 55–56 °C when the climatic chamber stayed at −34 °C. In particular, the ther-mal powers delivered by the heating gas lines placed upstream of the stack were not sufficient to keep temperatures of +60 °C at the stream inlets. Understandably, this trouble occurred especially when the reactive gas flow values were low (e.g., for gas amounts related to a 25-A current). The authors observed quite rapid decreases of the gas inlet temperatures (and stack temperature as well) when the lowest current values of the polarisation curve records were applied. Consequently, some small local con-densation/saturation could occur in the gas pipes. The authors emphasised that this problem could be aggravated if the fuel cell was required to operate at even lower tem-peratures (approximately −60 °C). In such a scenario, gas humidification would have to be reduced to avoid condensation in the gas pipes. This would lead to the drying of the stack membrane, resulting in a larger decay of the delivered power. In another test, 'operating low-temperature test', the stack was first purged using gas flow rates up to 150 N l/min and the stack and gas inlet temperatures were kept in the range +25 to +15 °C. The surrounding temperature was then fixed to –9 °C for 3 h. Once the stack temperature was uniformly close to –9 °C, reactive gas flows were set (14 and 55 N l/min for air and fuel, respectively) and the fuel cell was started up by increasing the load current from 0 to 25 A slowly over 15 min. After this, the surrounding chamber temperature was decreased to –18 °C. The stack achieved 1,500 W at 40 A, but failures in several cells and some anode–cathode leakages were detected. The authors attributed these to membrane perforations caused by the formation of ice inside the stack due to insufficient water draining out of the fuel cell during the previous drying procedure or due to excessive amounts of water produced by the reaction at sub-zero temperatures [87].

### 5.6.5 Transit bus application

On the city roads, buses go through regular acceleration and deceleration, which results in an extremely dynamic duty cycle. These conditions can give rise to fluctuations in the RH. Previously operated transit bus fleets [e.g., the fleet operated by Ballard using HD6 module in whistler resort community for 4 years (2010–2014)] have revealed mild reduction in humidity (≈10%) during acceleration, which is quickly recovered as the load reduces [88]. Although these short-duration changes are not likely to have affected the membrane shrinking or swelling, such recurring cycles would give rise to mechanical stress and fatigue leading to degradation. Apart from this, transit buses spend a lot of idle time due to stoppages at bus stops and traffic lights. These cause long periods of high voltages, which are known to trigger chemical degradation in the PEM. Considering these facts, Macauley and co-workers developed a specialised accelerated membrane durability test (AMDT) for heavy-duty fuel cells under bus-related conditions. They used a 10-cell stack with a non-reinforced PFSA membrane. GDE (with Pt/carbon and ionomer-coated catalyst layer) were used and the MEA was formed by hot pressing the GDE and membrane together. The stack was conditioned for 24 h before running the AMDT. The six AMDT and their specifications are listed in the Table 5.3. To accelerate degradation, these AMDT combined chemical and mechanical stressors. Furthermore, pure oxygen with a high partial pressure was used (transit buses mostly use air from the environment) to exacerbate radical formation and rapid degradation. To identify failure, the stack was considered to have failed when the internal leak rate of 100 sccm (i.e., the limiting leak rate of 10 sccm/cell through the membrane) as suggested by the US DOE was reached [89]. Additional *ex situ* mechanical tests were carried out to compare the BOL and end-of-test membranes.

**Table 5.3:** Details of the accelerated tests carried out by Macauley and co-workers for heavy-duty transit bus conditions.

| Test | Protocol conditions | Membrane lifetime (h) | Observations |
|------|---------------------|----------------------|--------------|
| Baseline | Stack voltage 9 V, RH cycling: RH drops to 60% every 10 min for 66 s at the cathode, OT: 85 °C, gas: 45% oxygen partial pressure, flow rates: $H_2$ at 5 slpm, $O_2$ at 10 slpm, current load ~1 A, back pressure 0.1 ± 0.3 barg | 298 | Fastest OCV decay, all 10 cells developed leaks Holes: diameter 50–300 µm, 2.9 holes/cm$^2$ OCV: voltage dropped to 8.3 V |
| Initiation | Baseline run until initial $H_2$ leaks detected *via* voltage fluctuations | 131 | Fastest OCV decay, all 10 cells developed leaks OCV: voltage dropped to 8.6 V Holes: diameter 200 µm, 1.3 holes/cm$^2$ Divot and crack formation at anode and cathode side |
| 90% RH | Baseline, but no RH cycling. Constant RH = 90% | 497 | Holes: diameter 40 µm, 0.04 holes/cm$^2$ |
| 100% RH | Baseline, but no RH cycling. Constant RH = 100% | 643 | Mildest condition – longest leak growth duration wrt 90% RH, 3 out of 10 membranes developed leaks during OCV OCV: dropped most severely to 8.0 V Holes: diameter 40 µm, 1.2 holes/cm$^2$ |
| PITM-1 | Platinum band generated using the Ballard protocol [7] prior to baseline | 405 | Pt band location: distance of 33% of the membrane thickness from the cathode interface, Pt concentration: 2,000– 18,000 ppm in the band OCV: voltage dropped to 8.6 V 1 divot, 0.68 holes/cm$^2$ |
| PITM-2 | Baseline+1 platinum band generating (using Ballard protocol) cycle every 6$^{th}$ RH cycle of baseline | 662 | Pt band location: 40% from the cathode Pt concentration: 11,000– 43,000 ppm 1 Sivot, 0.33 holes/cm$^2$ |

OT: Operating temperature
PITM: Platinum in the membrane
Reproduced with permission from N. Macauley, A.S. Alavijeh, M. Watson, J. Kolodziej, M. Lauritzen, S. Knights, G. Wang and E. Kjeang, *Journal of the Electrochemical Society*, 2015, **162**, 1, F98. ©2015, The Electrochemical Society [88]

According to the authors, the baseline AMDT allowed the lifetime of the membrane to be reduced by nearly one order of magnitude when compared with real bus operation, making the accelerated test very time-efficient.

The tests mentioned above revealed that the membrane lifetime increased significantly in the absence of any RH cycling (90 and 100% RH AMDT). However, nearly 150 h more of membrane performance were achieved with 100% RH as compared with 90% RH, which was attributed to the domination of the increase in oxygen partial pressure (due to reduced RH) over the reduced gas permeability, leading to oxygen crossover to the anode. On the other hand, RH cycling resulted in a gradual increase in the rate of degradation as opposed to a constant degradation rate in the case of 90% RH. The Pt band formation enhanced membrane durability, leading to minimal decay of OCV in the two PITM AMDT. The loss of membrane thickness, the fluoride ion release rates, and the SEM images of the degraded membranes after the various AMDT are shown in Figures 5.10 and 5.11. The *ex situ* tensile tests further demonstrated the enhanced durability of membranes due to Pt inside the membrane because only these PITM samples exhibited ductile behaviour similar to that of BOL samples (PITM-2 endured the tensile test, the PITM-1 sample fractured at ≈40% strain), whereas the AMDT samples fractured quickly beyond their yield stress. Mechanical tests at room temperature (23 °C, 50% RH) and fuel cell conditions (70 °C, 90% RH) showed PITM-1 to have lower elongation at fuel cell conditions, which could be an indication of some damage to the water channels inside the membrane. It was concluded the PITM-2, which showed no such variation, suffered no destruction of water channels due to a higher Pt concentration [89].

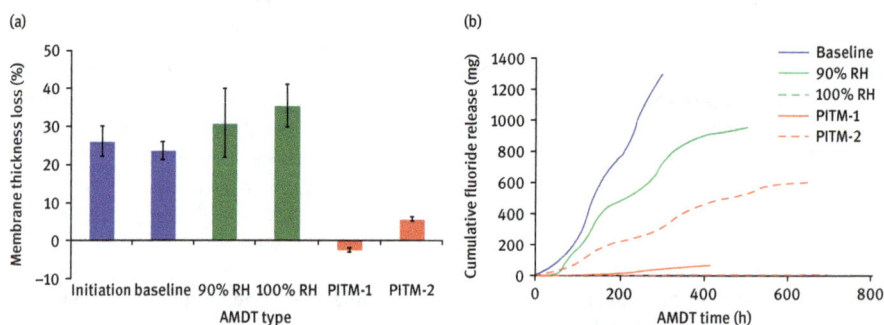

**Figure 5.10:** a) Membrane thickness loss, as percentage of original membrane thickness, measured using SEM at the end-of-life of after various AMDT and b) cumulative fluoride release during AMDT operations, obtained from conductivity measurements on the effluent water from the stacks. Reproduced with permission from N. Macauley, A.S. Alavijeh, M. Watson, J. Kolodziej, M. Lauritzen, S. Knights, G. Wang and E. Kjeang, *Journal of the Electrochemical Society*, 2015, **162**, 1, F98. ©2015, The Electrochemical Society [89].

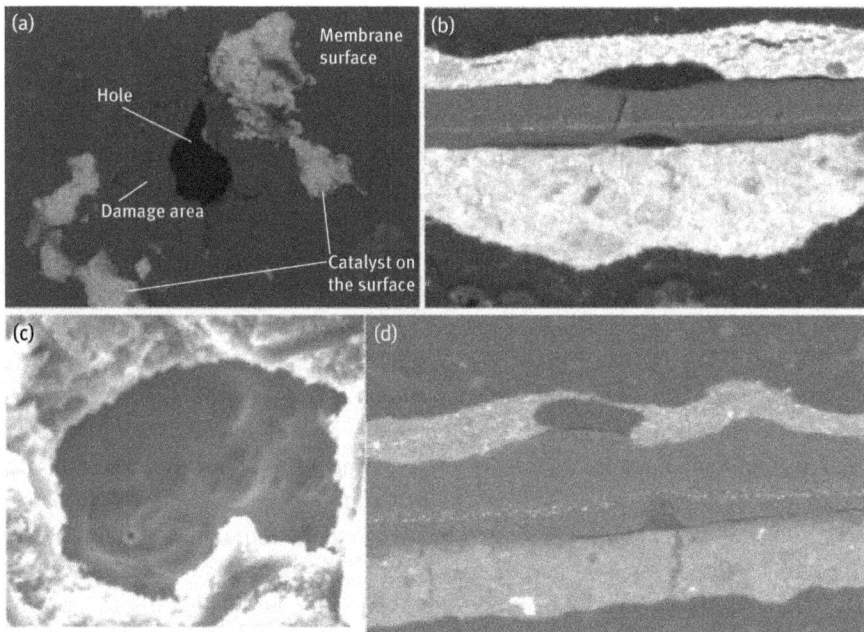

**Figure 5.11:** SEM images showing the membrane damage induced by AMDT: a) baseline; b) initiation; c) constant RH;and d) PITM AMDT. Reproduced with permission from N. Macauley, A.S. Alavijeh, M. Watson, J. Kolodziej,M. Lauritzen, S. Knights, G. Wang and E. Kjeang, *Journal of the Electrochemical Society*, 2015, **162**, 1, F98. ©2015,The Electrochemical Society [89].

## 5.7 Single-cell *in situ* testing of Nafion®-hybrid, other perfluorosulfonic acid and non-perfluorosulfonic acid membranes

This section discusses the various recent *in situ*, single-cell PEMFC and DMFC analyses reported on non-Nafion® and Nafion® hybrid and other membranes. Although only a limited number of studies are available so far, the trend of testing membranes other than Nafion® in real fuel cell setup is slowly picking up pace.

### 5.7.1 *In situ* proton-exchange membrane fuel cells testing at low and variable relative humidity conditions

A variety of PFSA composites (including Nafion® composites) and hydrocarbon-based membranes have been prepared and reported in recent years. The aim is to reduce the dependence of the PEM on humidity and water content so as to enhance operability and suitability under high temperature (HT) and low-RH conditions. Limited numbers of *in situ* studies on some of these PEM have also been seen over the last 2–3 years. Nafion® composite membranes with sulfonic acid-functionalised graphene (S-graphene) (1 wt%) as inorganic filler were prepared and tested *in situ* in a PEMFC single-cell setup at 70 °C and 20% RH in comparison with recast Nafion-® and a Nafion®–graphene composite by Sahu and co-workers. To carry out the *in situ* tests, catalyst loading of 0.5 mg/cm² was used on the anode and cathode where diffusion layer-coated carbon papers (SGL DC-35) were used as the electrodes. The active area for the PEMFC was 25 cm². MEA were prepared by placing the membrane between the cathode and anode and hot pressing under a pressure of 60 kg/cm² at 130 °C for 3 min. The PEM was tested by running the system at 100 °C and 20% RH, and parameters such as dew-point temperature, gas temperature/gas supply temperature, and dew-point humidification temperature were adjusted using the fuel cell test station to achieve RH values between 20 and 100%. For 100% RH, hot and wet hydrogen (450 ml/min) and oxygen (600 mL/min) were supplied to anode and cathode sides, respectively. According to the researchers, the high density of -SO$_3$H groups on S-graphene enabled higher IEC, leading S-graphene to behave as a solid-acid proton-conduction medium. Consequently, significantly higher peak power density and current density values could be achieved in comparison with Nafion® and Nafion®–graphene at 100% RH as well as in near-dry (20% RH) conditions. In addition, the S-graphene facilitated higher water adsorption and uptake, preventing the membrane from drying out at low RH and mitigating membrane flooding under high load current density [90].

Moving away from Nafion®, Lee and co-workers further explored the PEMFC application of polyether sulfones (PES) with clustered sulfonic groups at different RH (100, 80 and 53%) at 80 °C and compared the performances with Nafion® NR-212. The sulfonated polyether sulfone copolymers (S4PH-*x*-PS) consisted of four phenyl substituents at the 2,2′-, 6- and 6′-positions of 4,4′-diphenyl ether. Four polymers with varying mole% (*x*) of 4,4′-dihydroxy-2,2′,6,6′-tetraphenyldiphenyl ether in two diol monomers were prepared: S4PH-20-PS, S4PH-30-PS, S4PH-35-PS and S4PH-40-PS (Figure 5.12). For the PEMFC single-cell tests, highly pure hydrogen (200 sccm) and oxygen (500 sccm) were supplied to the anode and cathode, respectively. The back pressure was set to 1 atm. Single-cell tests at OCV revealed that S4PH-40-PS demonstrated the highest power density (462.6 mW/cm²), comparable with that of Nafion® NR-212 (533.5 mW/cm²) at 80 °C and 80% RH. However, S4PH- 40-PS failed at 100% RH due to poor mechanical strength. On the other hand, S4PH-20-PS showed poor performance at 80% RH but produced 362.5 mW/cm² under 100% RH at 80 °C, suggesting that ionic clusters within S4PH-20-PS failed to connect effectively under low RH. Further polarisation studies at 80% RH and 80 °C by the same research team identified S4PH-*x*-PS to have lower activation over potential (S4PH-30-PS, S4PH-35-PS < S4PH-40-PS) than that of Nafion® NR-212, leading to reduced current density. This led the authors to suggest that the excessive water uptake in

(a)

(b)

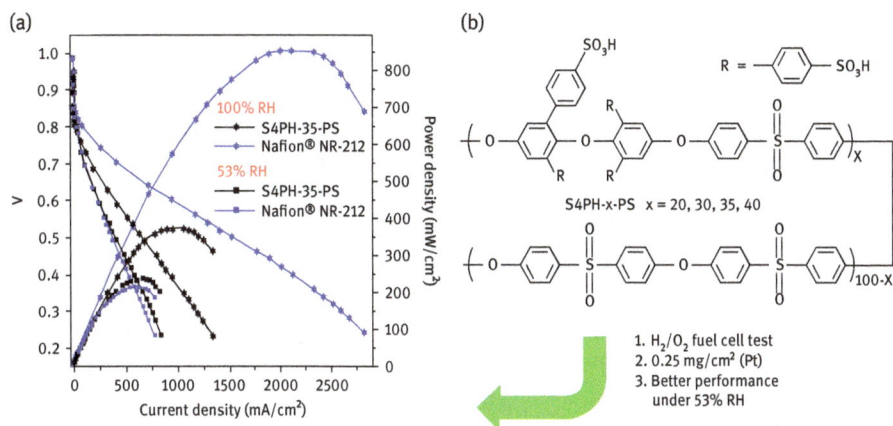

**Figure 5.12:** a) *In situ* PEMFC performance of Nafion® NR-212 and S4PH-35-PS at 100 and 80% RH and b) of S4PH-x-PS. Reproduced with permission from S. Lee, J. Chen, J. Wu and K. Chen, *ACS Applied Materials & Interfaces*, 2017, **9**, 11, 9805.©2017, American Chemical Society [91].

S4PH-40-PS could result in catalyst flooding and reduce the active surface area of Pt, thereby hindering oxygen transport in the catalyst layer. The role of the membrane–electrode interface and its negative influence on cell performance due to excessive swelling of S4PH-x-PS at higher RH was also highlighted. Further investigations at different RH established that the flooding effect of S4PH-35-PS on the catalyst layer decreased at low RH and, at 53% RH, 80 °C, fuel cells based on S4PH-35-PS displayed higher-peak power density (234.9 mW/cm²) than that of Nafion® NR-212 (214.0 mW/cm²) (Figure 5.12). The performance of the cells based on S4PH-35-PS and Nafion® NR-212 at different RH is summarised in Table 5.4 [91].

**Table 5.4:** Comparison of S4PH-35-PS and Nafion® NR-212 performance in single-cell PEMFC at different RH when operating at 80 °C as reported in the work by Lee [90].

| RH (%) | S4PH-35-PS | | | Nafion® NR-212 | | |
|---|---|---|---|---|---|---|
| | $PD_{peak}$ (mW/cm²) | $CD_{0.73\ V}$ (mA/cm²) | OCV (V) | $PD_{peak}$ (mW/cm²) | $CD_{0.73V}$ (mA/cm²) | OCV (V) |
| 53 | 234.9 | 68.6 | 0.900 | 214.0 | 88.1 | 0.988 |
| 66 | 289.1 | 52.8 | 0.838 | 344.7 | 118.1 | 0.927 |
| 80 | 445.1 | 124.1 | 0.930 | 533.5 | 206.6 | 0.958 |
| 100 | 372.9 | 117.3 | 0.922 | 847.4 | 255.0 | 0.953 |

$PD_{peak}$: Peak power density
$CD_{0.73V}$: Current density at 0.73 V
Reproduced with permission from S. Lee, J. Chen, J. Wu and K. Chen, *ACS Applied Materials & Interfaces*, 2017, **9**, 11, 9805. ©2017, American Chemical Society [90]

Poly(p-phenylene)-based aromatic hydrocarbon ionomers with a pendant PFSA group in a substituent at the 2-position have also shown promising performance at low RH conditions in *in situ* PEMFC tests [92]. Produced by a direct coupling polymerisation, these ionomers possess aromatic backbones without other linkages and have a high density of superacid groups on the flexible side chains, thus combining the positive features of perfluorinated and aromatic hydrocarbon ionomers. These membranes are relatively thin (18 μm) as compared with other reported hydrocarbon membranes (40–125 μm) and show excellent fuel cell performance under low humidity conditions at 80 °C with minimal loss of power density (907 mW/cm$^2$ at 30% RH and 976 mW/cm$^2$ at 90% RH). The authors reported that the MEA also showed high OCV of 0.98 V at 90% RH and 0.99 V at 30% RH due to the low gas permeability of the membranes.

Other factors affected by low RH operation have also been studied. Breitwieser and co-workers explored their direct membrane deposition (DMD) MEA systems for the effect of low RH on water management. The membrane is directly sprayed onto the anode and cathode GDE in such a setup. Consequently, the membranes are much thinner than those in traditional MEA setups, which is suggested to have benefits, such as easier humidification at the anode using back diffusion even at low humidity. In an *in situ* neutron radiography study using a 4-cm$^2$ active area cell with 12 μm-thick membrane, Breitwieser and co-workers studied water management at 70 °C under dry/15% and 35%/35% anode/cathode RH conditions. Their observations confirmed the presence of water on the non-humidified anode, substantiating facilitated back diffusion of reaction water from the cathode. This further explained the significantly improved fuel cell performance under dry conditions. Additional evidence (minimal increase in high-frequency resistance and identical water distribution in MEA as well as anode and cathode flow channels) supported the notion that, in the DMD fuel cell, back diffusion counterweighed the electro-osmotic drag even at higher current densities (>1,200 mA cm$^2$), ensuring sufficient anode humidification [93].

### 5.7.2 *In situ* direct methanol fuel cell testing of hybrid membranes

A variety of Nafion® composite/hybrids, non-Nafion® and even non-PFSA membranes have been studied for their performance in *in situ* single-cell DMFC. Although the tests reported so far are usually limited to I–V curves, they provide useful information about the suitability of these materials in real systems and form the foundation for more detailed *in situ* studies in the future.

Abouzari-Lotf and co-workers reported a single-cell DMFC study using a novel multilayer membrane consisting of phototungstic acid (PWA) self-immobilised on electrospun nanofibrous sheets of Nylon as the inner layer and Nafion® as the outer layer. Nafion® outer layers of 25–35 μm were prepared by casting 15% concentrated aqueous Nafion® solution. The PWA-anchored electrospun Nylon mats and the two outer Nafion® layers were mechanically compacted at 130°C (i.e., above the glass transition temperature of Nafion® and Nylon) and 1,560 psi for 30 s. Nafion® solution was sprayed on all three layers before hot pressing to enhance interlayer adhesion (Figure 5.13). For MEA preparation and single-cell testing, the membranes (pre-treated by boiling in 3 wt% $H_2O_2$ for 1 h, deionised water, 1 M $H_2SO_4$ and finally Milli-Q® water) were spray-coated using a catalyst ink consisting of electrocatalyst, 5 wt% Nafion® as a binder and water:isopropanol (1:1). Pt/ruthenium black (1:1) and Pt black were used as the anode and cathode electrocatalysts, respectively. Loadings of ≈4 mg/cm² were used for the anode and cathode. MEA (active area = 5 cm²) was prepared by hot pressing the CCM with carbon paper at 130 °C under 1,300 psi for 5 min. The DMFC test was done at 60 °C with 2 or 5 M methanol fed to the anode at a flow rate of 4 ml/min while humidified air (200 mL/min) was supplied to the cathode. The polarisation studies for the 3-layer (3L) system were done in comparison with Nafion® 115 (Figure 5.13). The 3L system demonstrated higher peak power, which was 49.6% and 113% higher than that of Nafion® with 2 M and 5 M methanol feed. A current density of 127.1 mW/cm² was observed for the 3L membrane as opposed to 59.6 mW/cm² for Nafion® at 5 M methanol. Moreover, the maximum power density of Nafion® 115-based MEA decreased by 15% upon increasing the methanol feed concentration, whereas it increased by 21.4% for 3L MEA. In addition, OCV, which is directly related to the methanol crossover, for the 3L system was ≈41 mV (for 2 M methanol) and 78 mV (for 5 M methanol) higher than that of Nafion® 115, further validating the superior methanol barrier properties of the 3L membrane system. The researchers also reported no leakage of PWA during testing [93]. Long-term AST with higher methanol feed for such promising systems could be very useful in developing DMFC with higher power and better durability.

In another study, Nafion®-free, polyvinyl alcohol (PVA)/sulfonated graphene oxide (SGO)/$Fe_3O_4$ membranes were prepared for use in DMFC. SGO sheets were decorated with iron oxide nanoparticles using the solvothermal method. In a unique process, the composite was then solution cast (solution of PVA and SGO/$Fe_3O_4$) in the presence of an applied magnetic field such that the SGO/Fe3O4 nanosheets were drawn to the through-plane direction of the cast membrane. Gluteraldehyde (GLA) was used as a crosslinking agent for PVA. For comparison studies, PVA/SGO/$Fe_3O_4$ with 0–7 wt% nanosheets along with other membranes (PVA/GLA, PVA/GLA/SGO cast in and out of a magnetic field) were also prepared. The as-prepared PVA/SGO/$Fe_3O_4$ membranes were reported to display higher PC, water uptake, thermal stability, methanol permeability, and selectivity when compared with a non-aligned membrane. *In situ* I–V tests were carried out on the membranes at 30 °C and 5% RH with 2 M aqueous methanol solution fed at a flow rate of 1 ml/min at the anode and $O_2$ fed at a flow rate of 300 ml/min at the cathode (area 3.5 × 3.5 cm). Catalyst loading of 2 mg/cm² Pt/ruthenium/carbon on the anode and 1 mg/cm² Pt/carbon on the cathode was used. *In situ* tests found that the PVA/SGO/$Fe_3O_4$ membrane cast inside a magnetic field with 5 wt% nanosheets ($M_{PSF-i5}$) demonstrated higher

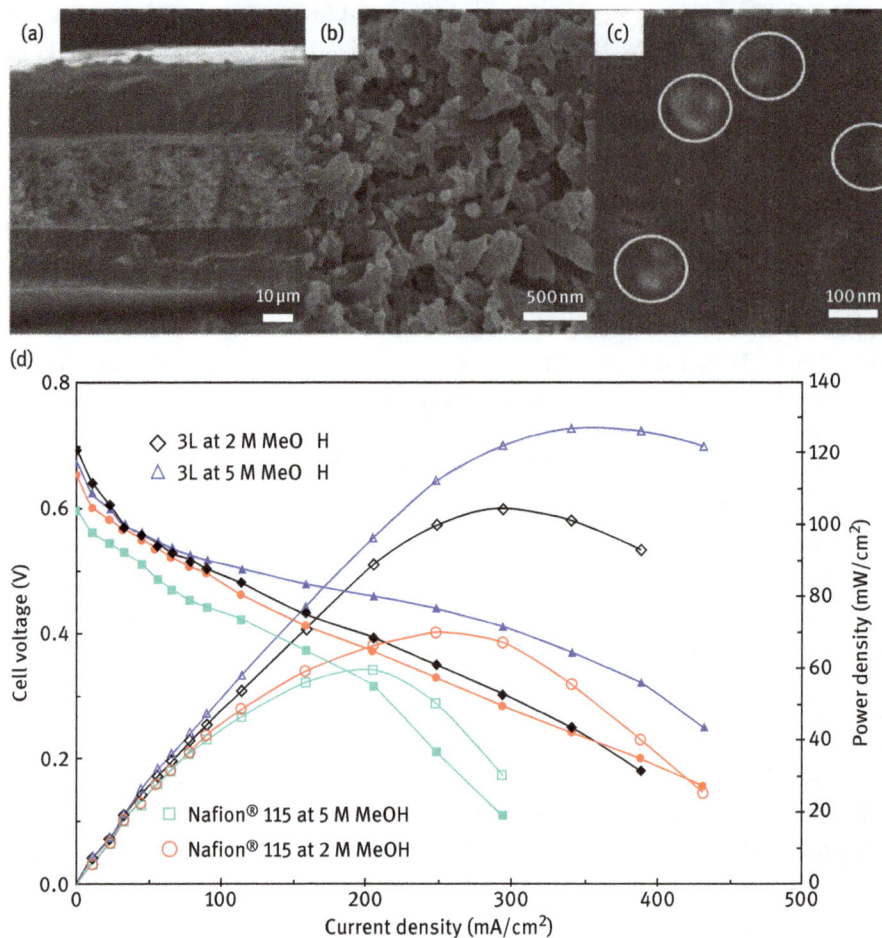

**Figure 5.13:** Field-emission SEM images showing (a) cross-section of Nafion®-PWA-anchored Nylon fibre multilayers; (b) zoomed in the central; (c) and top layers; and (d) polarisation and power curves of DMFC single cell with a Nafion® 115 membrane and 3L composite/multilayer membrane. Reproduced with permission from E. Abouzari-Lotf, M.M. Nasef, H. Ghassemi, M. Zakeri,A. Ahmad and Y. Abdollahi, *ACS Applied Materials & Interfaces*, 2015, **7**, 31, 17008. ©2015, American Chemical Society [94].

OCV (0.67 V) than that of $M_{PSF-o5}$ (cast outside a magnetic field) membrane (0.65 V). The maximum current density and power density of $M_{PSF-i5}$ was 107.11 mA/cm$^2$ and 25.57 mW/cm$^2$, respectively, which were higher than those of $M_{PSF-o5}$ (96.95 mA/cm$^2$ and 20.70 mW/cm$^2$) and were attributed to the enhanced water uptake and PC of $M_{PSF-i5}$. [95]. Feng and co-workers tested a semi-crystalline polyether ketone (SPEK)-based membrane. The semi-crystalline nature of the membrane was retained in dry and hydrated states, enabling low water uptake and low-volume swelling ratio, which further reduced methanol permeability. *In situ* tests reported by the research team showed that overall, semi-SPEK delivered better cell performances than amorphous sulfonated polyether ketone sulfone with higher OCV and less polarisation losses at 25 and 60 °C. However, it was pointed out that high polarisation loss was observed in samples with highly crystalline domains, which was attributed to the slightly low conductivity possibly due to the crystalline nature hindering water diffusion [96].

Nataraj and co-workers prepared a non-PFSA, natural polymer and clay composite, chitosan (CS)–montmorillonite (MMT) membrane. Composite membranes with 1, 3, 5 and 10 wt% sulfonated MMT were prepared using solution casting. MEA electrodes of Pt–ruthenium on carbon cloth (80 wt% 4.0 mg/cm²) and Pt black on carbon cloth (5.0 mg/cm²) were used as the anode and cathode, respectively. The electrodes with the CS–MMT were hot pressed at 130 °C using a force of 130 kg/cm² for 2 min. MEA performances were recorded under various methanol concentrations (2, 4, 6 and 8 M) where methanol solution was supplied into the anode at 20 mL/cm³ and $O_2$ flowed into the cathode (200 sccm) at 60 °C during all measurements. The researchers also carried out tests to observe methanol crossover by flowing $N_2$ into the cathode (200 sccm), whereas a positive voltage was applied to the cathode using a potentiostat (SI 1280Z, Solartron). In this method, the methanol at the cathode was oxidised to $CO_2$ under the applied voltage. At sufficiently high applied voltage, all methanol molecules are quickly oxidised at the cathode and the methanol crossover limiting current is reached, which is associated with the rate of methanol crossover at OCV. The authors reported an overall 3-fold reduction in methanol crossover as compared with Nafion® 117 (Figure 5.14). Their study also reported highly stable current density of ≈30 mW/cm² for the CS–MMT 3 wt% natural nanocomposite at high-concentration methanol feed (4, 6 and 8 M) (Figure 5.14) [97].

### 5.7.3 *In situ* intermediate- and high-temperature proton-exchange membrane fuel cell operation

Phosphoric acid (PA)-doped polybenzimidazole (PBI) and non-PBI membranes have garnered significant interest in recent years for HT-PEM application. PBI–PA acid–base membranes suffer from mechanical and thermal instability at high doping levels, which increases the risk of acid leaching. The acid–base pairs formed between the imidazole groups in the main chains of PBI and PA molecules result in a dynamic hydrogen bond network, which promotes proton transport. Consequently, high PA doping remains unavoidable if such membranes are to be used at 100–200 °C with low humidification or in anhydrous conditions without compromising on PC. To enhance the commercial suitability and balance the mechanical and proton-conduction properties of these acid–base systems, approaches such as introduction of micro-/nanostructured porous structures into the crosslinked or blended membrane composites which allow higher and regulated uptake

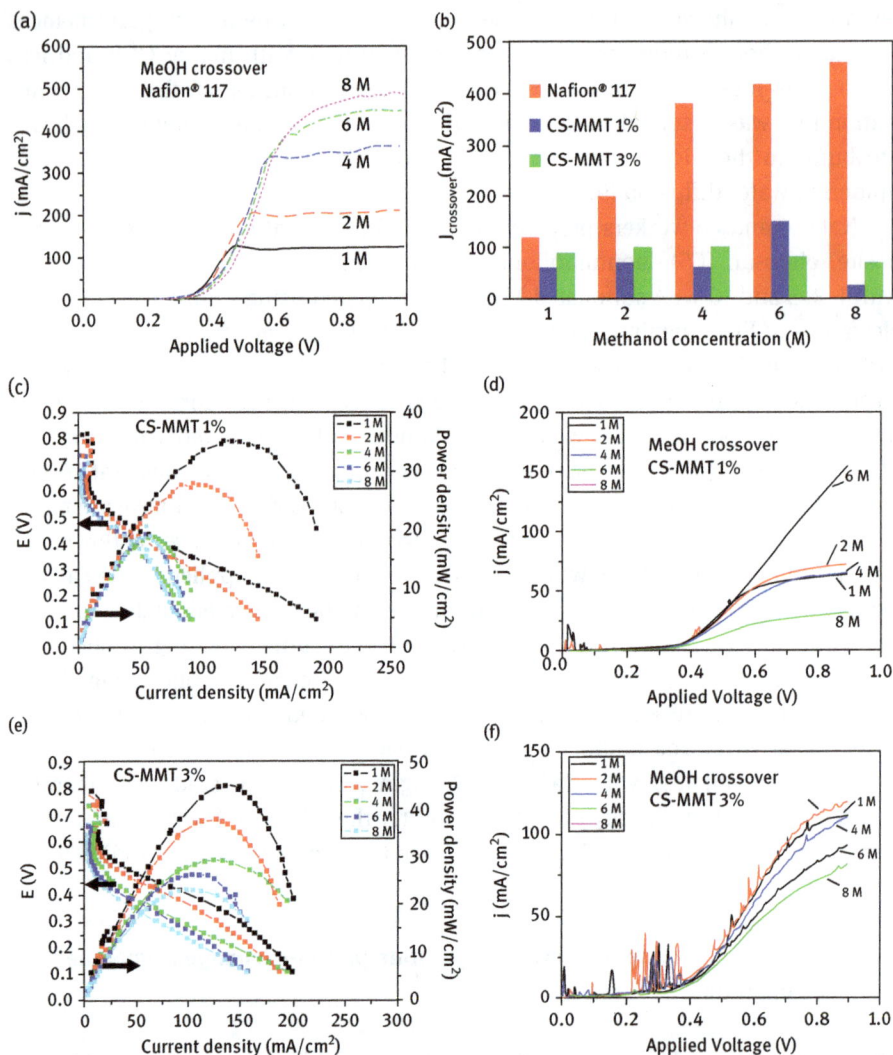

**Figure 5.14:** (a) Comparative measurement of methanol crossover current density (MCCD) as a function of applied voltage for commercial standard PEM Nafion®117 and (b) comparative representation of methanol-crossover limiting current densities determined by an applied voltage of 1 V of CS–MMT 1% and CS–MMT 3% with different methanol concentrations. (c) Polarisation curves of CS–MMT 1% and (d) its MCCD measurements (e) polarisation curves of CS–MMT 3% and (f) its MCCD measurements. Reproduced with permission from S.K. Nataraj, C. Wang, H. Huang, H. Du, L. Chen and K. Chen, *ACS Sustainable Chemistry & Engineering*, 2015, **3**, 2, 302.©2015, American Chemical Society [97].

of PA have been explored [98–104]. Guo and co-workers reported a PES–polyvinyl-pyrrolidone (PVP) membrane. Upon PA doping, the PA is permanently trapped inside the submicron-sized pores of this membrane. The excess acid (free acid) could lead to the creation of PA domains in the sub-microporous structure and facilitate transfer of protons *via* the Grotthuss mechanism, in accordance with pure PA. A connected sub-microporous structure was formed by monodispersing different wt% of $SiO_2$ solid spheres (diameter 250 nm) in the PES/PVP–dimethylformamide casting solution, followed by casting and hydrogen fluoride treatment. The membranes were tested in single-cell systems with Pt/carbon electrodes, (Pt loading = 0.5 mg/cm$^2$) on the anode and cathode, achieved by brushing catalyst ink onto the microporous GDL (dried at 45 °C). MEA was prepared by hot pressing, and single-cell tests were carried out at 120–180 °C with dry $H_2$ and $O_2$ (flow rates = 150 ml/min). OCV for cells operating with PES/PVP pores formed with 50 wt% $SiO_2$ (PES/PVP mp-50) was higher than that for pristine PES/PVP at 150 °C, suggesting reduced gas permeability. PES/PVP mp-50 and PES/PVP mp-30 also demonstrated excellent power density (218 mW/cm$^2$ for mp-30 and 361 mW/cm$^2$ for mp-50) in anhydrous conditions at 150 °C. The maximum power density reported was 454 mW/cm$^2$ at 180 °C under anhydrous conditions. The cell with PES/PVP mp-50 also did not reveal obvious degradation during the test over 150 h at a constant loading current of 0.2 A/cm$^2$ at 150 °C [104].

Zuo and co-workers reported single-cell studies on PVP–phosphonated poly (2,6-dimethyl-1,4-phenylene oxide) (pPPO) membranes with varying wt% of PVP and PVP–pPPO–graphitic carbon nitride nanocomposite membranes prepared by solution casting. Single-cell OCV tests at 180 °C and 0.33 V under anhydrous conditions revealed power density and current density of 138 mW/cm$^2$ and 420 mA/cm$^2$, respectively, for PA-doped 70%PVP/pPPO (with 70 wt% PVP). The inclusion of 5 wt% graphitic carbon nitride nanofillers into the 70%PVP/pPPO blend, which was primarily to improve the thermal and mechanical stability of the blend, also resulted in further performance enhancement, leading to maximum power density of 294 mW/cm$^2$ [105].

Zhang and co-workers investigated water transport and the effect of water vapour on cell performance when operating the cell at 160 °C with PBI/PA membrane in flow-through as well as in dead-end mode. Their cell consisted of a 5-step serpentine flow field graphite plate, 45 cm$^2$ active area, while dry $H_2$ and air (300 sccm) were supplied. Their investigations showed that when the cell was operated at 0.2 A/cm$^2$ in flow-through mode, no effect of water vapour could be seen on cell performance. However, in dead-end mode, the stability and performance of the cell was affected. They concluded that water vapour crossover could hinder $H_2$ diffusion in the anode, decreasing $H_2$ concentration and thus reducing cell performance. They also observed that with a fixed current load, the mass of accumulated water changed linearly, implying a constant rate of water transport at fixed current. Similar to low-temperature PEMFC, the water concentration gradient (which is proportional to the amount of current generated) is the main driving force for pushing water vapour from the cathode to the anode [106].

Lee and co-workers explored a composite PFSA membrane by impregnating Aquivion® with PA-modified titanium zirconium oxide ($TiO_2/ZrO_2$) nanofibres [107]. Aquivion® was chosen over Nafion® because of its better performance at >110 °C [108]. Single-cell studies of membranes consisting of 9 wt% uniformly distributed electrospun $TiO_2/ZrO_2$ (titanium/zirconium = 1:1 atomic ratio) were done in four testing conditions, which included fully-humidified medium-temperature conditions and partially humidified high-temperature conditions. Comparison studies were undertaken with Aquivion® only and a Aquivion®/$TiO_2/ZrO_2$ composite without PA treatment. The maximum power densities of the Aquivion®/$TiO_2/$ $ZrO_2$ composite membrane under 100% RH, low-temperature conditions (1,120 and 917 $mW/cm^2$ at 75 and 90 °C, respectively) were higher than those of the Aquivion® (695 and 694 $mW/cm^2$) membrane. However, at higher temperature and low RH conditions, the Aquivion®-only membrane performed better than the composites ($\approx$362 at 120 °C/40% RH and 167 $mW/cm^2$ at 140 °C/20% RH). On the other hand, the Aquivion® composite impregnated with PA-modified $TiO_2/ZrO_2$ demonstrated good performance under all conditions (Pmax = 1.18 $W/cm^2$ at 75 ° C/100% RH, and 0.97 $W/cm^2$ at 90 °C/100% RH 0.45 $W/cm^2$ at 120 °C/40% RH, and 0.21 $W/cm^2$ at 140 °C/20% RH). The high water retention of the Aquivion®/ phosphate-modified $TiO_2/ZrO_2$ nanofibre composite membrane assisted better performance under fully-humidified conditions, while the phosphate functionality increased PC by boosting proton transfer *via* the Grotthuss mechanism at higher temperature and lower RH conditions. Furthermore, the accelerated lifetime test (ALT) undertaken on the phosphate-treated composite and other membranes revealed a lower reduction rate for the current density of the phosphate-treated composite (−8.57 $mA/cm^2/h$) as opposed to Aquivion® membrane (−11.74 $mA/$ $cm^2/h$), which faced a drastic loss of current density after 15 h. The enhanced cell durability of the phosphate-modified composite was attributed to thermally- and mechanically-stable inorganic fibres and phosphate functionality, which resulted in superior PC at HT and low RH. Impedance spectra of the membranes carried out before and after the ALT showed that the phosphate-modified membrane exhibited lower charge transfer resistance than the Aquivion® membrane, both before and after the ALT (Figure 5.15). When examining LSV scans for $H_2$ crossover, the phosphate-modified membrane also showed a minimum change in current density (from 0 to 0.21 $mA/cm^2$) even after the ALT whereas, in the case of Aquivion®, it increased rapidly (current density increased from 2.22 to 6.78 $mA/cm^2$). All recorded current density values were <10 $mA/cm^2$, so no obvious $H_2$ crossover through the membranes was reported but some potential membrane damage (in the Aquivion®-only membrane) associated with material degradation or pinhole formation could not be ruled out.

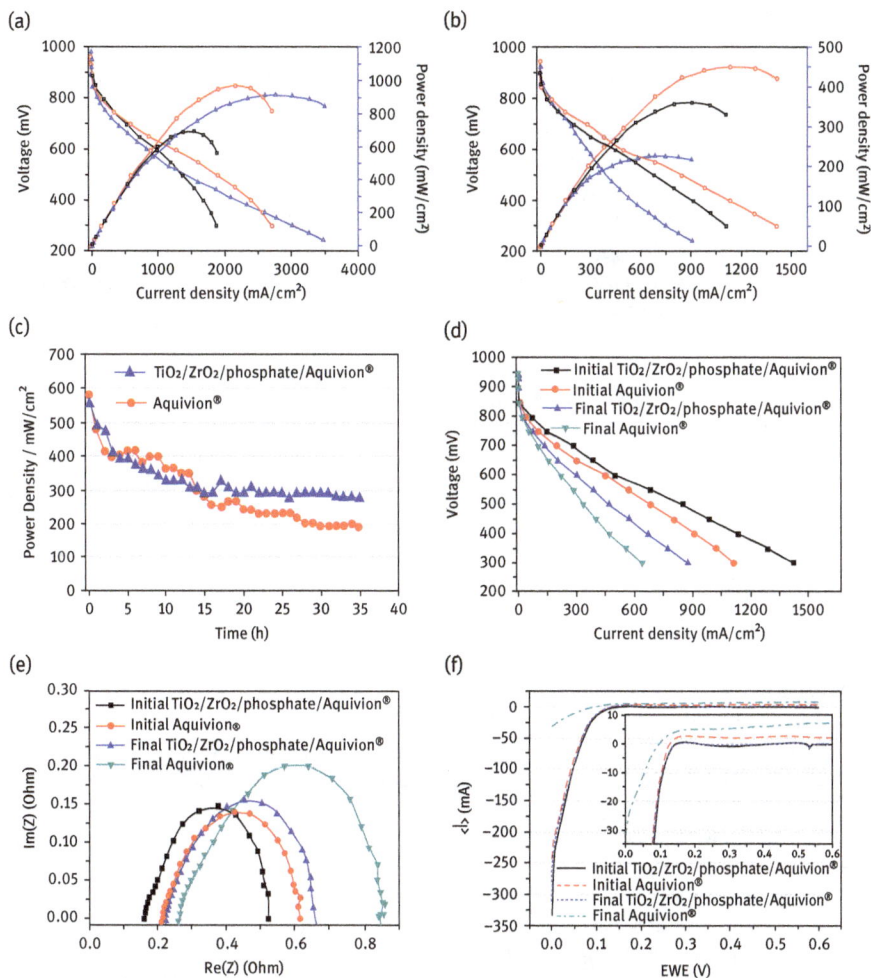

**Figure 5.15:** Polarisation and power density curves of phosphate- modified $TiO_2/ZrO_2$ nanofibre/ Aquivion®, $TiO_2/ZrO_2$ nanofibre/ Aquivion®, and Aquivion® membranes at different testing conditions. (a) 90 °C and 100% RH and, (b) 120 °C and 40% RH, (c) variation of the current density of $TiO_2/ZrO_2$/phosphate/ Aquivion® and Aquivion® during the ALT at 0.6 V under 120 °C, 40% RH, and atmospheric pressure, and initial and final electrochemical analysis: (d) polarisation curve, (e) impedance spectra, and (f) LSV curve. Reproduced with permission from C. Lee, J.H. Park, Y. Jeon, J. Park, H. Einaga, Y.B. Truong, I.L. Kyratzis, I. Mochida, J. Choi and Y.G. Shul, *Energy & Fuels*, 2017, **31**, 7645. ©2017, American Chemical Society [107].

# References

[1] J. Wu, X.Z. Yuan, J.J. Martin, H. Wang, D. Yang, J. Qiao and J. Ma, *Journal of Power Sources*, 2010, **195**, 4, 1171.

[2] H. Zhang and P.K. Shen, *Chemical Reviews*, 2012, **112**, 5, 2780.

[3] M.P. Rodgers, L.J. Bonville, H.R. Kunz, D.K. Slattery and J.M. Fenton, *Chemical Reviews*, 2012, **112**, 11, 6075.

[4] M. Zhao, W. Shi, B. Wu, W. Liu, J. Liu, D. Xing, Y. Yao, Z. Hou, P. Ming and Z. Zou, *Electrochimica Acta*, 2015, **153**, 254.

[5] R. Banan, A. Bazylak and J. Zu, *International Journal of Hydrogen Energy*, 2015, **40**, 4, 1911.

[6] N.S. Khattra, A.M. Karlsson, M.H. Santare, P. Walsh and F.C. Busby, *Journal of Power Sources*, 2012, **214**, 365.

[7] C. Lim, L. Ghassemzadeh, F. Van Hove, M. Lauritzen, J. Kolodziej, G. Wang, S. Holdcroft and E. Kjeang, *Journal of Power Sources*, 2014, **257**, 102.

[8] M. Fumagalli, S. Lyonnard, G. Prajapati, Q. Berrod, L. Porcar, A. Guillermo and G. Gebel, *The Journal of Physical Chemistry B*, 2015, **119**, 23, 7068.

[9] P.W. Majsztrik, M.B. Satterfield, A.B. Bocarsly and J.B. Benziger, *Journal of Membrane Science*, 2007, **301**, 1, 93.

[10] M.B. Satterfield and J. Benziger, *The Journal of Physical Chemistry B*, 2008, **112**, 12, 3693.

[11] D.T. Hallinan, Jr., and Y.A. Elabd, *The Journal of Physical Chemistry B*, 2009, **113**, 13, 4257.

[12] D.T. Hallinan, Jr., M.G. De Angelis, M. Giacinti Baschetti, G.C. Sarti and Y.A. Elabd, *Macromolecules*, 2010, **43**, 10, 4667.

[13] A. Kusoglu, S. Savagatrup, K.T. Clark and A.Z. Weber, *Macromolecules*, 2012, **45**, 18, 7467.

[14] A. Rollet, G. Gebel, J. Simonin and P. Turq, *Journal of Polymer Science, Part B: Polymer Physics*, 2001, **39**, 5, 548.

[15] L. Maldonado, J.C. Perrin, J. Dillet and O. Lottin, *Journal of Membrane Science*, 2012, **389**, 43.

[16] A. El Kaddouri, J. Perrin, T. Colinart, C. Moyne, S. Leclerc, L. Guendouz and O. Lottin, *Macromolecules*, 2016, **49**, 19, 7296.

[17] T. Colinart, J. Perrin and C. Moyne, *Journal of Polymer Science, Part B: Polymer Physics*, 2014, **52**, 22, 1496.

[18] T.V. Nguyen and R.E. White, *Journal of the Electrochemical Society*, 1993, **140**, 8, 2178.

[19] D.M. Bernardi, *Journal of the Electrochemical Society*, 1990, **137**, 11, 3344.

[20] D.M. Bernardi and M.W. Verbrugge, *AIChE Journal*, 1991, **37**, 8, 1151.

[21] D.M. Bernardi and M.W. Verbrugge, *Journal of the Electrochemical Society*, 1992, **139**, 9, 2477.

[22] S. Motupally, A.J. Becker and J.W. Weidner, *Journal of the Electrochemical Society*, 2000, **147**, 9, 3171.

[23] T.A. Zawodzinski, Jr., M. Neeman, L.O. Sillerud and S. Gottesfeld, *The Journal of Physical Chemistry*, 1991, **95**, 15, 6040.

[24] T.F. Fuller in *Solid-Polymer-Electrolyte Fuel Cells*, California University, Berkeley, Lawrence Berkeley Lab, Berkeley, CA, USA, 1992, p.79. [PhD Thesis]

[25] Q. Yan, H. Toghiani and J. Wu, *Journal of Power Sources*, 2006, **158**, 1, 316.

[26] S. Tsushima, K. Teranishi and S. Hirai, *Electrochemical and Solid-State Letters*, 2004, **7**, 9, A269.

[27] J.P. Owejan, J.J. Gagliardo, J.M. Sergi, S.G. Kandlikar and T.A. Trabold, *International Journal of Hydrogen Energy*, 2009, **34**, 8, 3436.

[28] J. Park, X. Li, D. Tran, T. Abdel-Baset, D.S. Hussey, D.L. Jacobson and M. Arif, *International Journal of Hydrogen Energy*, 2008, **33**, 13, 3373.

[29] J.H. Nam and M. Kaviany, *International Journal of Heat and Mass Transfer*, 2003, **46**, 24, 4595.

[30]  H. Ju, C. Wang, S. Cleghorn and U. Beuscher, *Journal of the Electrochemical Society*, 2005, **152**, 8, A1645.

[31]  Y. Lee, B. Kim and Y. Kim, *International Journal of Hydrogen Energy*, 2009, **34**, 18, 7768.

[32]  Y. Tabuchi, R. Ito, S. Tsushima, S. Hirai, A. Horai, K. Aotani, N. Kubo and K. Shinohara, *ECS Transactions*, 2010, **33**, 1, 1045.

[33]  P. Huguet, A. Morin, G. Gebel, S. Deabate, A. Sutor and Z. Peng, *Electrochemistry Communications*, 2011, **13**, 5, 418.

[34]  K. Kunimatsu, B. Bae, K. Miyatake, H. Uchida and M. Watanabe, *The Journal of Physical Chemistry B*, 2011, **115**, 15, 4315.

[35]  K. Kunimatsu, T. Yoda, D. Tryk, H. Uchida and M. Watanabe, *Physical Chemistry Chemical Physics*, 2010, **12**, 3, 621.

[36]  M. Hara, J. Inukai, K. Miyatake, H. Uchida and M. Watanabe, *Electrochimica Acta*, 2011, **58**, 1, 449.

[37]  F.I. Allen, L.R. Comolli, A. Kusoglu, M.A. Modestino, A.M. Minor and A.Z. Weber, *ACS Macro Letters*, 2014, **4**, 1, 1.

[38]  M. Falk, *Canadian Journal of Chemistry*, 1980, **58**, 14, 1495.

[39]  S. Paddison, *Journal of New Materials for Electrochemical Systems*, 2001, **4**, 4, 197.

[40]  Y. Choe, E. Tsuchida, T. Ikeshoji, A. Ohira and K. Kidena, *The Journal of Physical Chemistry B*, 2010, **114**, 7, 2411.

[41]  M. Schalenbach, W. Lueke, W. Lehnert and D. Stolten, *Electrochimica Acta*, 2016, **214**, 362.

[42]  M. Khandelwal and M.M. Mench, *Journal of Power Sources*, 2006, **161**, 2, 1106.

[43]  C. Choy, Y. Wong, G. Yang and T. Kanamoto, *Journal of Polymer Science, Part B: Polymer Physics*, 1999, **37**, 23,3359.

[44]  K.A. Mauritz and R.B. Moore, *Chemical Reviews*, 2004, **104**, 10, 4535.

[45]  M.H. Eikerling and P. Berg, *Soft Matter*, 2011, **7**, 13, 5976.

[46]  H. Ito, T. Maeda, A. Nakano and H. Takenaka, *International Journal of Hydrogen Energy*, 2011, **36**, 17, 10527.

[47]  M. Schalenbach, T. Hoefner, P. Paciok, M. Carmo, W. Lueke and D. Stolten, *The Journal of Physical Chemistry C*, 2015, **119**, 45, 25145.

[48]  C. Rogers, *Polymer Permeability*, 1985, **2**, 11.

[49]  T. Sakai, H. Takenaka, N. Wakabayashi, Y. Kawami and E. Torikai, *Journal of the Electrochemical Society*, 1985, **132**, 6, 1328.

[50]  T. Dorfmueller, W.T. Hering, K. Stierstadt and G. Fischer in *Lehrbuch der Experimental Physik*, Volume 1, 11ᵗʰ Edition, Ed., W. De Gruyter, Berlin, Germany and New York, NY, USA, 1998, p.902

[51]  K. Kreuer, *Solid State Ionics*, 2013, **252**, 93.

[52]  M. Goulet, S. Arbour, M. Lauritzen and E. Kjeang, *Journal of Power Sources*, 2015, **274**, 94

[53]  S. Kundu, M. Fowler, L. Simon and S. Grot, *Journal of Power Sources*, 2006, **157**, 2, 650.

[54]  C. Lim, L. Ghassemzadeh, F. Van Hove, M. Lauritzen, J. Kolodziej, G. Wang, S. Holdcroft and E. Kjeang, *Journal of Power Sources*, 2014, **257**, 102.

[55]  A. Tavassoli, C. Lim, J. Kolodziej, M. Lauritzen, S. Knights, G.G. Wang and E. Kjeang, *Journal of Power Sources*, 2016, **322**, 17.

[56]  A. Jung, I.M. Kong, C.Y. Yun and M.S. Kim, *Journal of Membrane Science*, 2017, **523**, 138.

[57]  A.S. Alavijeh R.M.H. Khorasany, Z. Nunn, A. Habisch, M. Lauritzen, E. Rogers, G.G. Wang and E. Kjeang, *Journal of Electrochemical Society*, 2015, **162**, 14, F1461.

[58]  Y. Li, D.A. Dillard, Y. Lai, S.W. Case, M.W. Ellis, M.K. Budinski and C.S. Gittleman, *Journal of the Electrochemical Society*, 2011, **159**, 2, B173.

[59]  R.M. Khorasany, E. Kjeang, G. Wang and R. Rajapakse, *Journal of Power Sources*, 2015, **279**, 55.

[60] *Fuel Cell Technologies Office Multi-Year Research, Development, and Demonstration Plan*, US Department of Energy, Washington, DC, USA. https://energy.gov/eere/fuelcells/downloads/ fuel-cell-technologies-office-multi-year-research-development-and-22 [Last accessed May 2017].

[61] Z. Wan, H. Chang, S. Shu, Y. Wang and H. Tang, *Energies*, 2014, **7**, 5, 3179.

[62] H. Markötter, I. Manke, R. Kuhn, T. Arlt, N. Kardjilov, M.P. Hentschel, A. Kupsch, A. Lange, C. Hartnig and J. Scholta, *Journal of Power Sources*, 2012, **219**, 120.

[63] E.L. Thompson, T. Capehart, T.J. Fuller and J. Jorne, *Journal of the Electrochemical Society*, 2006, **153**, 12, A2351.

[64] M. Cappadonia, J.W. Erning and U. Stimming, *Journal of Electroanalytical Chemistry*, 1994, **376**, 1–2, 189.

[65] M. Cappadonia, J.W. Erning, S.M.S. Niaki and U. Stimming, *Solid State Ionics*, 1995, **77**, 65.

[66] H. Yoshida and Y. Miura, *Journal of Membrane Science*, 1992, **68**, 1–2, 1.

[67] A. Siu, J. Schmeisser and S. Holdcroft, *The Journal of Physical Chemistry B*, 2006, **110**, 12, 6072.

[68] M. Pineri, G. Gebel, R.J. Davies and O. Diat, *Journal of Power Sources*, 2007, **172**, 2, 587.

[69] H. Mendil-Jakani, R.J. Davies, E. Dubard, A. Guillermo and G. Gebel, *Journal of Membrane Science*, 2011, **369**, 1, 148.

[70] J. Biesdorf, P. Stahl, M. Siegwart, T.J. Schmidt and P. Boillat, *Journal of the Electrochemical Society*, 2015, **162**, 10, F1231.

[71] M.A. Hickner, H. Ghassemi, Y.S. Kim, B.R. Einsla, J.E. Mcgrath, M.A. Hickner, H. Ghassemi, Y.S. Kim, B.R. Einsla and J.E. Mcgrath, *Chemical Reviews*, 2004, **104**, 10, 4587.

[72] A. Roudgar, S. Narasimachary and M. Eikerling, *The Journal of Physical Chemistry B*, 2006, **110**, 41, 20469.

[73] R.C. McDonald, C.K. Mittelsteadt and E.L. Thompson, *Fuel Cells*, 2004, **4**, 3, 208.

[74] E. Cho, J. Ko, H.Y. Ha, S. Hong, K. Lee, T. Lim and I. Oh, *Journal of the Electrochemical Society*, 2004, **151**, 5, A661.

[75] Y. Oh, S.K. Kim, D.H. Peck, D.H. Jung and Y. Shul, *International Journal of Hydrogen Energy*, 2014, **39**, 28, 15760.

[76] S. Zhang, X. Yuan, R. Hiesgen, K.A. Friedrich, H. Wang, M. Schulze, A. Haug and H. Li, *Journal of Power Sources*, 2012, **205**, 290.

[77] G. Diloyan, M. Sobel, K. Das and P. Hutapea, *Journal of Power Sources*, 2012, **214**, 59.

[78] G. Watts and V. Krylov, *Applied Acoustics*, 2000, **59**, 3, 221.

[79] R. Hassan and K. McManus, *Journal of Low Frequency Noise, Vibration and Active Control*, 2002, **21**, 2, 65.

[80] Ş Yildirim, S. Erkaya, İ Eski and İ Uzmay, *Journal of Vibration and Control*, 2009, **15**, 1, 133.

[81] M.C. Betournay, G. Bonnell, E. Edwardson, D. Paktunc, A. Kaufman and A.T. Lomma, *Journal of Power Sources*, 2004, **134**, 1, 80.

[82] V. Rouss, P. Lesage, S. Bégot, D. Candusso, W. Charon, F. Harel, X. François, V. Selinger, C. Schilo and S. Yde-Andersen, *International Journal of Hydrogen Energy*, 2008, **33**, 22, 6755.

[83] V. Rouss, D. Candusso and W. Charon, *International Journal of Hydrogen Energy*, 2008, **33**, 21, 6281.

[84] N. Rajalakshmi, S. Pandian and K. Dhathathreyan, *International Journal of Hydrogen Energy*, 2009, **34**, 9, 3833.

[85] Y. Hou, D. Hao, J. Shen, P. Li, T. Zhang and H. Wang, *International Journal of Hydrogen Energy*, 2016, **41**, 9, 5123.

[86] R. Banan, A. Bazylak and J. Zu, *International Journal of Hydrogen Energy*, 2013, **38**, 34, 14764.

[87] S. Bégot, F. Harel, D. Candusso, X. François, M.C. Péra and S. Yde-Andersen, *Energy Conversion and Management*, 2010, **51**, 7, 1522.

[88]   *Component Accelerated Stress Test Protcols for PEM Fuel Cells*, USCAR Fuel Cell Tech Team, US Department of Energy, Washington, DC, USA, 2010. http://www1.eereenergy.gov/hydroge nandfuelcells/pdfs/component_durability_may_2010.pdf [last accessed July 2017].

[89]   N. Macauley, A.S. Alavijeh, M. Watson, J. Kolodziej, M. Lauritzen, S. Knights, G. Wang and E. Kjeang, *Journal of the Electrochemical Society*, 2015, **162**, 1, F98.

[90]   A.K. Sahu, K. Ketpang, S. Shanmugam, O. Kwon, S. Lee and H. Kim, *The Journal of Physical Chemistry C*, 2016, **120**, 29, 15855.

[91]   S. Lee, J. Chen, J. Wu and K. Chen, *ACS Applied Materials & Interfaces*, 2017, **9**, 11, 9805.

[92]   T. Oshima, M. Yoshizawa-Fujita, Y. Takeoka and M. Rikukawa, *ACS Omega*, 2016, **1**, 5, 939.

[93]   M. Breitwieser, R. Moroni, J. Schock, M. Schulz, B. Schillinger, F. Pfeiffer, R. Zengerle and S. Thiele, *International Journal of Hydrogen Energy*, 2016, **41**, 26, 11412.

[94]   E. Abouzari-Lotf, M.M. Nasef, H. Ghassemi, M. Zakeri, A. Ahmad and Y. Abdollahi, *ACS Applied Materials & Interfaces*, 2015, **7**, 31, 17008.

[95]   H. Beydaghi and M. Javanbakht, *Industrial & Engineering Chemistry Research*, 2015, **54**, 28, 7028.

[96]   S. Feng, J. Pang, X. Yu, G. Wang and A. Manthiram, *ACS Applied Materials & Interfaces*, 2017, **9**, 29, 24527.

[97]   S.K. Nataraj, C. Wang, H. Huang, H. Du, L. Chen and K. Chen, *ACS Sustainable Chemistry & Engineering*, 2015, **3**, 2, 302.

[98]   J. Wang, H. Yu, M. Lee, S. Zhang and D. Wang, *Journal of Applied Polymer Science*, 2012, **124**, 4, 3175.

[99]   M. Hazarika and T. Jana, *ACS Applied Materials & Interfaces*, 2012, **4**, 10, 5256.

[100]  Y. Chang, J. Lai and Y. Liu, *Journal of Membrane Science*, 2012, **403**, 1.

[101]  C. Shen, L. Jheng, S.L. Hsu and J.T. Wang, *Journal of Materials Chemistry*, 2011, **21**, 39, 15660.

[102]  Q. Tang, G. Qian and K. Huang, *RSC Advances*, 2013, **3**, 11, 3520.

[103]  S. Wang, C. Zhao, W. Ma, G. Zhang, Z. Liu, J. Ni, M. Li, N. Zhang and H. Na, *Journal of Membrane Science*, 2012, **411**, 54.

[104]  Z. Guo, R. Xiu, S. Lu, X. Xu, S. Yang and Y. Xiang, *Journal of Materials Chemistry A*, 2015, **3**, 16, 8847.

[105]  G. Zou, W. Wu, C. Cong, X. Meng, K. Zhao and Q. Zhou, *RSC Advances*, 2016, **6**, 108, 106237.

[106]  C. Zhang, L. Zhang, W. Zhou, Y. Wang and S.H. Chan, *Electrochimica Acta*, 2014, **149**, 271.

[107]  C. Lee, J.H. Park, Y. Jeon, J. Park, H. Einaga, Y.B. Truong, I.L. Kyratzis, I. Mochida, J. Choi and Y.G. Shul, *Energy & Fuels*, 2017, **31**, 7645.

[108]  A. Stassi, I. Gatto, E. Passalacqua, V. Antonucci, A. Arico, L. Merlo, C. Oldani and E. Pagano, *Journal of Power Sources*, 2011, **196**, 21, 8925.

# 6 Future trends in polymer electrolyte membranes

## 6.1 Introduction

Among the vast 'sea' of novel and dynamic research concepts explored in polymer electrolyte membranes (PEM), some have gained more credibility over the years in terms of their future potential. Concepts such as self-humidification, anhydrous proton conduction, multilayered membranes and three-dimensional printed membranes offer considerable potential for the development of thinner, more durable, excellent proton conduction, high-performing PEM with minimal-to-zero hydration requirements. This concluding chapter briefly touches upon these concepts, exploring their scope and future perspectives before discussing the recyclability of PEM.

## 6.2 Future trends towards next-generation membranes: Enhancing proton conduction, durability and minimising the dependence on humidification

Development of membranes with better water retention or self-hydration mechanisms that can suffice membrane hydration requirements based on the water generated at the cathode will broaden the range of operational temperature and humidity for the PEM, and enhance the overall performance and durability of fuel cell systems. Moreover, these membranes can also minimise (or completely bypass) the need for external humidification systems, water management systems, which significantly add to the overall mass and size of the fuel cell peripherals.

Initial reports on self-humidifying membranes were published in the early-2000s. Since then, several sulfonated and hydrocarbon composites of this type have been investigated [1–4]. However, this concept is in its early stages and there is immense scope for growth in this area. Earlier studies explored the growth/incorporation of a layer/band of platinum or platinum/carbon particles into Nafion®, and tried to understand the effect of the location (distance from the anode and cathode) of this band. Platinum is unstable at high voltages, so it dissolves and migrates out of the catalyst layer, and is deposited in the membrane. The partial pressure of crossover hydrogen ($H_2$) is pertinent for platinum band formation because $H_2$ reacts with platinum ions, reducing them to metallic platinum inside the membrane. The local mixed potential distribution inside the membrane defines the location of the platinum band [5–9]. More recent investigations have revealed that generation/formation of the platinum band enables self-humidification if located near the anode catalyst layer through water generated as a result of crossover gases reacting at the platinum-electroactive sites. It also results in a higher open circuit voltage and enhances the mechanical stability and durability of the membrane by

https://doi.org/10.1515/9783110647327-006

ensuring the membrane does not dry up under dry conditions and consuming the gases, which would otherwise contribute to membrane degradation [10–16]. Nevertheless, the effect of platinum in the membrane remains controversial because counter-arguments regarding platinum ions behaving as promoters of Fenton's reactions are supported by strong experimental evidence [17, 18].

Over the years, other membrane materials (sulfonated polymers such as sulfonated polyether ether ketone) as well as $Cs_{2.5}H_{0.5}PW_{12}O_{40}$-supported platinum have been explored [19]. These restrict gas crossover while enhancing proton conductivity (PC) at fuel cell operating conditions. Initiatives to move away form platinum have led to exploration of other metals (gold, silver, palladium) for formation of metallic bands inside the PEM. Such initiatives have revealed dramatic reduction in the fluoride emission rate (also referred to as the 'fluoride release rate') during accelerated stress tests on membrane electrode assemblies (MEA) [20]. The recognition of $CeO_2$, $ZrO_2$ and transition metals (chromium, cobalt, manganese) as scavengers of hydroxyl radicals, which can mitigate the membrane degradation, have led to further exploration into this multi-pronged approach. Combination of this concept (incorporation of metal/transition metal bands) with various fabrication methods, including direct membrane deposition and silica-supported transition metal-based radical scavengers integrated into the short side chains of perfluorosulfonic acid (PFSA) membranes, are being explored [21–25].

Another approach towards minimising the dependence on water is that of anhydrous proton conduction. Anhydrous systems are of high interest for proton-exchange membrane fuel cells (PEMFC) operating at 100–200 °C. Metal organic frameworks (MOF) and porous coordination polymers have garnered interest as anhydrous, solid-state proton conductors because they can encapsulate mobile proton carriers (such as $H_3PO_4$, $H_2PO_4^-$, $H_2O$ and amines) due to their consistent and tailorable pore structures and thermal stability. They, however, face drawbacks such as leaching and fuel crossover [26–33]. To overcome these challenges, a recent study exploited the defect sites in non-porous coordination polymer crystals for encapsulating proton carriers. Defects introduced into a monodentate $H_2PO_4^-$ network can be used to embed uncoordinated $H_3PO_4$ as proton carriers, and they seem to retain the carriers without leakage up to the decomposition temperatures of the polymer crystals and allow high PC (with reference to existing proton-conduction polymers, MOF, acid-doped polymer systems) at 30–150 °C. This unique approach has potential for application to other proton-conductive crystalline solids, such as MOF, metal phosphates, and oxoacids (e.g., $CsHSO_4$) in the future [34].

Polymer ionic liquid(s) (PIL)-based membranes also offer anhydrous proton conduction along with high thermal, chemical as well as electrochemical stability. PIL are well-suited for intermediate temperature proton-exchange membrane fuel cells (IT-PEMFC) and high-temperature proton-exchange membrane fuel cell operations. The challenge in the case of ionic liquids (IL) is to immobilise the room temperature ionic liquid (RTIL) into a solid state for material application. This is overcome by PILs, which retain the qualities of RTIL while acquiring a more suit-able solid structure. Incorporation of PIL into polybenzimidazole (PBI) membranes is attracting huge attention. Different methods/approaches for preparation of unique porous-PBI structures and inclusion of PIL into these microstructures to achieve high-performing membranes are being explored continually [35–39]. Recently, Kallem and co-workers explored a unique high-temperature polymer elec-trolyte membrane concept consisting of hierarchically structured PIL channels em-bedded in an ordered porous PBI [hierarchical polybenzimidazole (HPBI)] microsieve. The liquid-induced phase-separation micromoulding method they em-ployed allowed manufacture of 30–40 μm-thick HPBI microsieves with 42–46% po-rosity and a definitive pore architecture (Figure 6.1). The inner surface created by the intrinsic porosity of HPBI facilitated IL confinement and enabled additional pathways for proton transport through the PIL network, allowing an operating tem-perature of 200 °C in anhydrous conditions [40].

## 6.3 Disposal and recycling of Ionomers and other membranes

Given the motivations of green energy, environmental concerns, and climate change riding on the back of fuel cell technology along with the economic costs as-sociated with their components, recycling and appropriate disposal of the materials used inside a fuel cell are of paramount importance. Recycling to recover/reclaim expensive and rare platinum/metal electrocatalysts is of major interest but atten-tion must also be paid towards the recycling of the expensive PEM used in the MEA. MEA incineration to recover the more precious metallic content would seem to be a straightforward option, but it is not cost-effective because the membranes add sub-stantially to the cost of the MEA. Moreover, small electrocatalyst metal particles tend to diffuse into and get trapped within the membranes, and incineration would decrease the recycling efficiency and make it more complicated [41]. In addition, fluorine-based membranes are recognised as contaminants in bipolar plates. Plati-num recycling *via* incineration or energy recycling as the possible formation of hydrogen fluoride at increased temperatures cannot be ignored, which in turn would need expensive hydrogen fluoride recycling plants [42]. Hydrocarbon and PBI/phosphoric acid membranes would be relatively easy to recycle or incinerate by conventional methods as opposed to PFSA membranes. Studies suggest that for a 60-kg MEA input, a 19-kg ionomer can be retrieved, which has a total worth of $8–20,000 depending on evaluation of the recovered polymer. A market for recy-cling and recovery of these polymer materials is currently lacking, but the situation is expected to improve with the expansion of the fuel cell industry and market [43]. Understandably, most studies focus on the recovery and recycling of platinum and few studies have investigated membrane-recycling options. Nonetheless, the poten-tial to recover full membranes for reuse in other applications is promising. Chemical

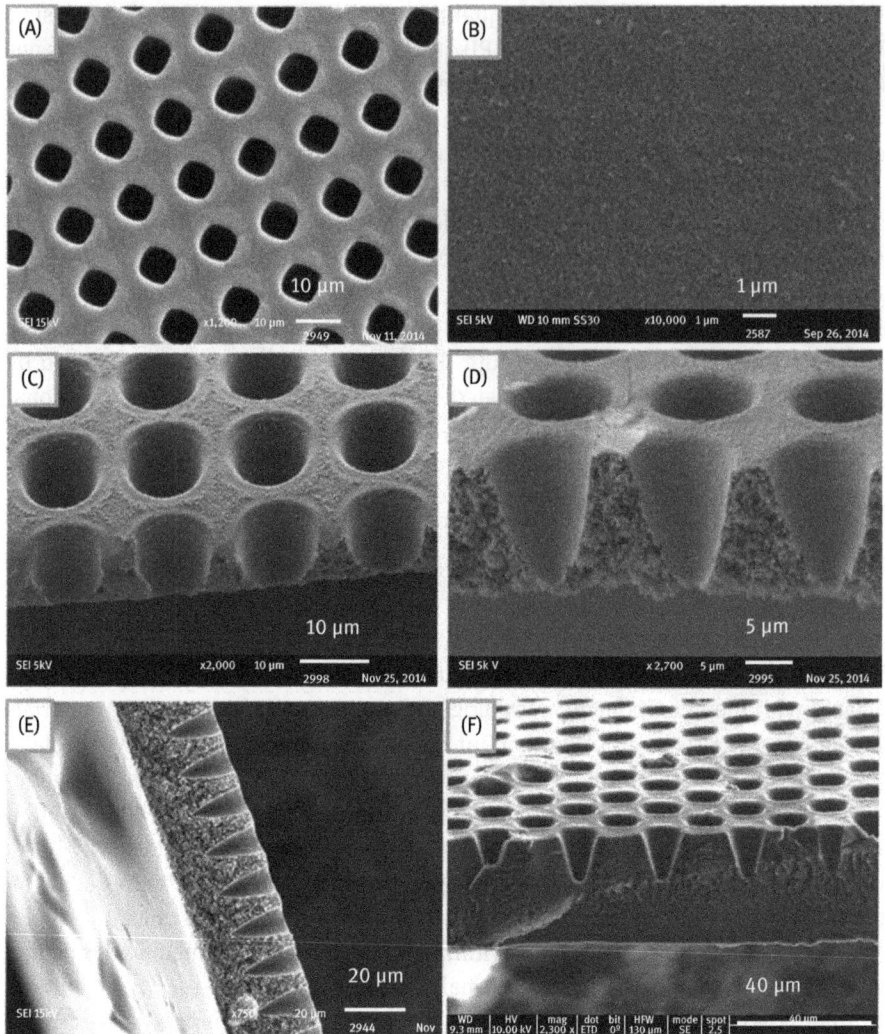

**Figure 6.1:** Scanning electron microscopy images of an HPBI microsieve with different skin layer thicknesses: (A) P-25 µm surface mould side; (B) P-25 µm surface air side; (C) P-3 µm cross-section; (D) P-5 µm cross-section; (E) P-15 µm cross-section; and (F) P-25 µm cross-section. Reproduced with permission from P. Kallem, M. Drobek, A. Julbe, E.J. Vriezekolk, R. Mallada and M.P. Pina, *ACS Applied Materials & Interfaces*, 2017, **9**, 17, 14844. ©2017, American Chemical Society [40].

dissolution or physical-separation methods are the two most common approaches to membrane recovery from MEA. Processes have been developed for recycling of MEA wherein the polymer material's performance is essentially recovered to the initial level after the removal of foreign cations inserted into the membrane during fuel cell operations [43]. Thermal and supercritical or ultrasonic treatments for membrane separation followed by further hydrometallurgical processing are the most efficient ways to retrieve the membrane [41, 44, 45].

Another approach could be to use MEA prepared without hot pressing. Some studies have suggested that non-hot pressed MEA perform better, and reach optimum performance quicker. Such MEA would make membrane and catalyst recovery easier for recycling and re-use [46].

Conversely, the use of less expensive natural polymers or biodegradable materials for proton-conduction membranes could allow for more convenient recycling or recovery (or even elimination).

## 6.4 Conclusions

With increasing numbers of studies being published on PEMFC and studies investigating fuel cell operations above 80 or 100 °C, there is a clear trend for moving towards IT-PEMFC.

An expansive selection of materials has been explored as membrane materials for low-temperature fuel cells. A wide variety of polymeric materials has managed to seize most of the attention, but others, such as carbon nanomaterials [carbon nanotubes, graphene oxide (GO), sulfonated GO, carboxylated GO], solid acids, clays and natural polymers (polysaccharides and proteins) have also generated significant research interest over the last 15 years. These materials bring in strong options and prospects for realisation of the demands [self-humidifying, relative humidity (RH)-independent proton conduction, thinner yet mechanically strong, impermeable to gas, and methanol transport] of the next generation of fuel cell membranes. Continued experimentation on combining these materials using appropriate methods and identifying the optimum proportions will provide insight about the vast potential and permutations of properties these combinations can offer. PFSA composites as well as other polymer composites and multilayer membranes are showing promising prospects towards development of self-humidifying, durable structures. Although most of the self-humidifying membranes are in developmental stages or relatively expensive compared with Nafion® and other commercial membranes, the costs are likely to decrease with the development of more efficient membrane processing and scale-up from laboratory to pilot studies.

Another conceivable approach in the future is the biomimetic approach, which uses synthetic materials to mimic and generate structures formed in nature. Bogdanowicz and co-workers reported the use of side-chain liquid crystalline polyethers dendronised with potassium 3,4,5-*tris*[4-(n-dodecan-1-yloxy)benzyloxy]benzoate to form oriented and stable membranes. When cast as membranes, these polymers self-assemble into columns. The process mimics the highly-supramolecular self-organisation seen in nature (such as the proteins in the tobacco mosaic virus), which is driven by exo-recognition. The polymer columns along the membrane length seem to form ionic paths delivering conductivity ($10^{-2}$–$10^{-3}$ S/cm, comparable with Nafion®) independent of RH [47].

Apart from these materials, membrane preparation methods can significantly impact the behaviour and properties of the membrane material formed into a membrane. Extruded (including reinforced) membranes continue to be the commercial standard, but various other new and innovative initiatives are being explored towards improvising the preparation methods to maximise the potential of membrane materials. These can further pave the way towards enhancing catalyst efficiency with thinner membranes without compromising on other membrane characteristics (durability, methanol/gas permeability), reducing the fuel cell size, and making the MEA more flexible to allow the penetration of low-temperature fuel cells into more electronic devices. Concepts of multilayered membranes, hybrid MEA and biomimetics demonstrate immense potential for revolutionising the very core (MEA) of existing fuel cell systems by paving the way towards lighter, thinner and highly efficient flexible (yet sturdy) systems which can blend low-temperature fuel cell systems beyond handheld electronics into wearable electronics.

With the development of physical and chemical characterisation and analytical methods, novel membranes and membrane materials can be investigated extensively. These methods help in the development of each type of membrane before the membrane is deemed fit or worthy of *in situ* fuel cell examination. However, systematic studies using standard and comparable test procedures for various temperature and humidity conditions must be carried out when studying new membranes. These studies can serve as a unifying platform providing detailed, analogous profiles of the different membranes for effective comparisons. *Ex situ* studies using RH cycling to study fracture and fatigue failure have revealed the causes of crack formation and their typical locations in polymer membranes, and also validated the notions of crack propagation even in the absence of geometric or electrochemical effects [48–51]. A gas-phase Fenton's test has validated the combined chemo-mechanical effects on degradation [52]. Accelerated stress tests such as Fenton's test, RH cycling and stress–strain/mechanical cycling provide valuable information. *In situ* test systems elicit an environment in which the membrane faces the real, dynamic challenges associated with the fuel cell. A wide variety of accelerated stress tests are utilised regularly for deeper understanding of the functioning of membranes as well as other components in specialised, simulated environments that mimic the extreme conditions and repetitive, cyclic fatigues that membranes would be subjected to during their lifetime. This has led to a substantial increase in our knowledge about PFSA, PBI and hydrocarbon membranes.

Nevertheless, the demand for better, cheaper, and eco-friendly membranes and fuel cell systems continues to drive research into membranes. This has led to a plethora of novel and unique membrane materials and concepts that are reported in literature continuously. These new materials and membrane systems incessantly stimulate fresh challenges requiring vigorous *ex situ* and *in situ* testing. Though *ex situ* testing in various difficult environments for various new membranes has shown promise, *in situ* performances are very much awaited.

Lastly, from recycling and ecological perspectives, further research on fluorine free/non-PFSA membrane fuel cell systems with lower amounts of fluorine in the membrane is required because these would be more eco-friendly for recycling and green disposal. Combination of natural materials with nanostructured carbons, metal oxides and clays also demands further experimental exploration to fully evaluate the possibilities of natural polymers which, apart from being eco-friendly and biodegradable, can offer high economic viability.

# References

[1]   F. Liu, B. Yi, D. Xing, J. Yu, Z. Hou and Y. Fu, *Journal of Power Sources*, 2003, **124**, 1, 81.
[2]   E. Chalkova, M.B. Pague, M.V. Fedkin, D.J. Wesolowski and S.N. Lvov, *Journal of the Electrochemical Society*, 2005, **152**, 6, A1035.
[3]   X. Zhu, H. Zhang, Y. Zhang, Y. Liang, X. Wang and B. Yi, *The Journal of Physical Chemistry B*, 2006, **110**, 29, 14240.
[4]   K.W. Feindel, S.H. Bergens and R.E. Wasylishen, *Journal of the American Chemical Society*, 2006, **128**, 43, 14192.
[5]   W. Gu, R.N. Carter, T.Y. Paul and H.A. Gasteiger, *ECS Transactions*, 2007, **11**, 1, 963.
[6]   F.N. Büchi, M. Inaba and T.J. Schmidt in *Polymer Electrolyte Fuel Cell Durability*, Eds., F.N. Büchi, M. Inaba and T.J. Schmidt, Springer, Berlin, Germany, 2009.
[7]   L. Dubau, J. Durst, F. Maillard, M. Chatenet, J. André and E. Rossinot, *Fuel Cells*, 2012, **12**, 2, 188.
[8]   S. Burlatsky, M. Gummalla, V. Atrazhev, D. Dmitriev, N. Kuzminyh and N. Erikhman, *Journal of the Electrochemical Society*, 2011, **158**, 3, B322.
[9]   H. Schulenburg, B. Schwanitz, J. Krbanjevic, N. Linse, G.G. Scherer and A. Wokaun, *Electrochemistry Communications*, 2011, **13**, 9, 921.
[10]  N. Macauley, L. Ghassemzadeh, C. Lim, M. Watson, J. Kolodziej, M. Lauritzen, S. Holdcroft and E. Kjeang, *ECS Electrochemistry Letters*, 2013, **2**, 4, F33.
[11]  O.T. Holton and J.W. Stevenson, *Platinum Metals Review*, 2013, **57**, 4, 259.
[12]  S.F. Burlatsky, J.B. Hertzberg, N.E. Cipollini, D.A. Condit, T.D. Jarvi, J.A. Leistra, M.L. Perry and T.H. Madden, inventors and assignee; US0224216A1, 2004.
[13]  N. Macauley, A.S. Alavijeh, M. Watson, J. Kolodziej, M. Lauritzen, S. Knights, G. Wang and E. Kjeang, *Journal of the Electrochemical Society*, 2015, **162**, 1, F98.
[14]  N. Macauley, L. Ghassemzadeh, C. Lim, M. Watson, J. Kolodziej, M. Lauritzen, S. Holdcroft and E. Kjeang, *ECS Electrochemistry Letters*, 2013, **2**, 4, F33.
[15]  M. Aoki, H. Uchida and M. Watanabe, *Electrochemistry Communications*, 2006, **8**, 9, 1509.
[16]  N. Macauley, K.H. Wong, M. Watson and E. Kjeang, *Journal of Power Sources*, 2015, **299**, 139.
[17]  E. Guilminot, A. Corcella, M. Chatenet, F. Maillard, F. Charlot, G. Berthomé, C. Iojoiu, J. Sanchez, E. Rossinot and E. Claude, *Journal of the Electrochemical Society*, 2007, **154**, 11, B1106.
[18]  C. Iojoiu, E. Guilminot, F. Maillard, M. Chatenet, J. Sanchez, E. Claude and E. Rossinot, *Journal of the Electrochemical Society*, 2007, **154**, 11, B1115.
[19]  P. Sayadi, S. Rowshanzamir and M.J. Parnian, *Energy*, 2016, **94**, 292.

[20] P. Trogadas, J. Parrondo, F. Mijangos and V. Ramani, *Journal of Materials Chemistry*, 2011, **21**, 48, 19381.

[21] P. Trogadas, J. Parrondo and V. Ramani, *Chemical Communications*, 2011, **47**, 41, 11549.

[22] Z. Wang, H. Tang, H. Zhang, M. Lei, R. Chen, P. Xiao and M. Pan, *Journal of Membrane Science*, 2012, **421–422**, 201.

[23] M. Breitwieser, T. Bayer, A. B\uchler, R. Zengerle, S.M. Lyth and S. Thiele, *Journal of Power Sources*, 2017, **351**, 145.

[24] C. D'Urso, C. Oldani, V. Baglio, L. Merlo and A. Aricò, *Journal of Power Sources*, 2016, **301**, 317.

[25] M.T. Taghizadeh and M. Vatanparast, *Journal of Colloid and Interface Science*, 2016, **483**, 1.

[26] S. Horike, D. Umeyama and S. Kitagawa, *Accounts of Chemical Research*, 2013, **46**, 11, 2376.

[27] T. Yamada, K. Otsubo, R. Makiura and H. Kitagawa, *Chemical Society Reviews*, 2013, **42**, 16, 6655.

[28] M. Yoon, K. Suh, S. Natarajan and K. Kim, *Angewandte Chemie International Edition*, 2013, **52**, 10, 2688.

[29] P. Ramaswamy, N.E. Wong and G.K. Shimizu, *Chemical Society Reviews*, 2014, **43**, 16, 5913.

[30] S.C. Sahoo, T. Kundu and R. Banerjee, *Journal of the American Chemical Society*, 2011, **133**, 44, 17950.

[31] N.C. Jeong, B. Samanta, C.Y. Lee, O.K. Farha and J.T. Hupp, *Journal of the American Chemical Society*, 2011, **134**, 1, 51.

[32] J.M. Taylor, K.W. Dawson and G.K. Shimizu, *Journal of the American Chemical Society*, 2013, **135**, 4, 1193.

[33] H.A. Patel, N. Mansor, S. Gadipelli, D.J. Brett and Z. Guo, *ACS Applied Materials & Interfaces*, 2016, **8**, 45, 30687.

[34] M. Inukai, S. Horike, T. Itakura, R. Shinozaki, N. Ogiwara, D. Umeyama, S. Nagarkar, Y. Nishiyama, M. Malon, A. Hayashi, T. Ohhara, R. Kiyanagi and S. Kitagawa, *Journal of the American Chemical Society*, 2016, **138**, 27, 8505.

[35] P. Kallem, A. Eguizabal, R. Mallada and M.P. Pina, *ACS Applied Materials & Interfaces*, 2016, **8**, 51, 35377.

[36] J. Lemus, A. Eguizábal and M. Pina, *International Journal of Hydrogen Energy*, 2015, **40**, 15, 5416.

[37] A.S. Shaplov, R. Marcilla and D. Mecerreyes, *Electrochimica Acta*, 2015, **175**, 18.

[38] J. Lemus, A. Eguizabal and M. Pina, *International Journal of Hydrogen Energy*, 2016, **41**, 6, 3981.

[39] M. Díaz, A. Ortiz, M. Vilas, E. Tojo and I. Ortiz, *International Journal of Hydrogen Energy*, 2014, **39**, 8,3970.

[40] P. Kallem, M. Drobek, A. Julbe, E.J. Vriezekolk, R. Mallada and M.P. Pina, *ACS Applied Materials & Interfaces*, 2017, **9**, 17, 14844.

[41] R. Wittstock, A. Pehlken and M. Wark, *Recycling*, 2016, **1**, 3, 343.

[42] C. Handley, N. Brandon and R. Van Der Vorst, *Journal of Power Sources*, 2002, **106**, 1, 344.

[43] J. Cooper in *Polymer Electrolyte Membrane and Direct Methanol Fuel Cell Technology: Volume 1: Fundamentals and Performance of Low Temperature Fuel Cells*, Eds., C. Hartnig and C. Roth,Woodhead, Cambridge, UK, 2012, p.117.

[44] T. Oki, T. Katsumata, K. Hashimoto and M. Kobayashi, *Materials Transactions*, 2009, **50**, 7, 1864.

[45] L. Shore, inventor; BASF Corp., assignee; US8124261, 2012.

[46] C. Song and P. Pickup, *Journal of Applied Electrochemistry*, 2004, **34**, 10, 1065.

[47] K.A. Bogdanowicz, S.V. Bhosale, Y. Li, I.F. Vankelecom, R. Garcia-Valls, J.A. Reina and M. Giamberini, *Journal of Membrane Science*, 2016, **509**, 10.

[48] A. Kusoglu, M.H. Santare and A.M. Karlsson, *Journal of Polymer Science, Part B: Polymer Physics*, 2011, **49**, 21, 1506.

[49] K. Patankar, D.A. Dillard, S.W. Case, M.W. Ellis, Y. Li, Y. Lai, M.K. Budinski and C.S. Gittleman, *Journal of Polymer Science, Part B: Polymer Physics*, 2010, **48**, 3, 333.

[50] Y. Li, D.A. Dillard, S.W. Case, M.W. Ellis, Y. Lai, C.S. Gittleman and D.P. Miller, *Journal of Power Sources*, 2009, **194**, 2, 873.

[51] H. Tang, S. Peikang, S.P. Jiang, F. Wang and M. Pan, *Journal of Power Sources*, 2007, **170**, 1, 85.

[52] W. Yoon and X. Huang, *ECS Transactions*, 2010, **33**, 1, 907.

# Abbreviations

| | |
|---|---|
| 1D | One-dimensional |
| 2D | Two-dimensional |
| AC | Alternating current |
| AFC | Alkaline fuel cell(s) |
| ALT | Accelerated lifetime test |
| AMDT | Accelerated membrane durability tests |
| AMST | Accelerated mechanical stress tests |
| AST | Accelerated stress test |
| ASTM | American Society for Testing and Materials |
| ATR | Attenuated total reflection |
| BOL | Beginning-of-life |
| BPP | Bipolar plate(s) |
| CCM | Catalyst-coated membranes |
| CE | Counter electrode |
| CNC | Cellulose nanocrystal(s) |
| CNF | Cellulose nanofibre(s) |
| CNT | Carbon nanotube(s) |
| CO | Carbon monoxide |
| COCV | Cyclic open circuit voltage |
| CS | Chitosan |
| DC | Direct current |
| DGMS | Direct gas mass spectrometry |
| $D_m$ | Mutual diffusion coefficient |
| DMD | Direct membrane deposition |
| DMFC | Direct methanol fuel cells |
| DOE | Department of Energy |
| DS | Degree of sulfonation |
| $D_s$ | Self-diffusion coefficient |
| DSC | Differential scanning calorimetry |
| $E_A$ | Activation energy(ies) |
| ECSA | Electrochemically-active surface area |
| EIS | Electrochemical impedance spectroscopy |
| EW | Equivalent weight |
| F/T | Freeze/thaw |
| FEP | Poly(fluoroethylene-*co*-hexafluoropropylene) |
| FTIR | Fourier-Transform infrared |
| GDE | Gas diffusion electrode |
| GDL | Gas diffusion layers |
| GE | General Electric |
| GLA | Glutaraldehyde |
| GO | Graphene oxide |
| HEP | Hexafluoropropylene |
| HPBI | Hierarchical polybenzimidazole |
| HT | High temperature |
| ICE | Internal combustion engine |
| IEC | Ion-exchange capacity |
| IL | Ionic liquid(s) |

https://doi.org/10.1515/9783110647327-007

| | |
|---|---|
| IR | Infrared |
| IT | Intermediate temperature |
| IT-PEMFC | Intermediate temperature proton-exchange membrane fuel cell(s) |
| LSV | Linear sweep voltammetry |
| LTA | Linde Type A |
| MCCD | Methanol crossover current density |
| MEA | Membrane electrode assembly(ies) |
| MMT | Montmorillonite |
| MOF | Metal organic frameworks |
| MOR | Mordenite |
| MS | Mass spectroscopy |
| MWCNT | Multi-walled carbon nanotube(s) |
| Na–MMT | Sodium montmorillonite |
| NMR | Nuclear magnetic resonance |
| NPI | Naphthalenic polyimide |
| OCV | Open circuit voltage |
| OT | Operating temperature(s) |
| P4VP | Poly(4-vinylpyrrolidone) |
| PA | Phosphoric acid |
| PAA | Polyacrylic acid |
| PAFC | Phosphoric acid fuel cell(s) |
| PAH | Polyallylamine hydrochloride |
| PBI | Polybenzimidazole |
| PC | Proton conductivity |
| PCM | Proton-conduction membrane |
| PDDA | Polydiallyl dimethyl ammonium chloride |
| PEI | Polyetherimide |
| PEM | Polymer electrolyte membrane |
| PEMFC | Proton-exchange membrane fuel cell(s) |
| PEO | Polyethylene oxide |
| PES | Polyether sulfone(s) |
| PFA | Poly(tetrafluoroethylene-*co*-perfluorovinyl ether) |
| PFCA | Perfluorocycloalkene |
| PFCI | Perfluorocarboxylated ionomer |
| PFGNMR | Pulsed-field gradient nuclear magnetic resonance |
| PFSA | Perfluorosulfonic acid |
| PIL | Polymer ionic liquid |
| PITM | Platinum in the membrane |
| pPPO | Phosphonated poly(2,6-dimethyl-1,4-phenylene oxide) |
| PSEPVE | Perfluorosulfonyl fluoride ethyl propyl vinyl ether |
| PSS | Polyphenylsulfone |
| PSSA | Polystyrene sulfonic acid |
| Pt | Platinum |
| PTFE | Polytetrafluoroethylene |
| PVA | Polyvinyl alcohol |
| PVDF | Polyvinylidene fluoride |
| PVP | Polyvinylpyrrolidone |
| PVT | Poly(1-vinyl-1,2,4-triazole) |
| PWA | Phototungstic acid |

| | |
|---|---|
| RH | Relative humidity |
| RT | Room temperature |
| RTIL | Room temperature ionic liquid |
| S4PH-*x*-PS | Sulfonated polyether sulfone copolymers |
| SANS | Small-angle neutron scattering |
| SEM | Scanning electron microscopy |
| SGO | Sulfonated graphene oxide |
| S-graphene | Sulfonic acid-functionalised graphene |
| SOFC | Solid oxide fuel cell(s) |
| SPAEK | Sulfonated poly aryl ether ketone |
| SPEEK | Sulfonated polyether ether ketone |
| SPEK | Semi-crystalline polyether ketone |
| SPES | Sulfonated polyether sulfone |
| SPPBP | Sulfonated poly(phenoxy benzoyl phenylene) |
| SPSU | Sulfonated polysulfone |
| TAP | Triaminopyrimidine |
| TEM | Transmission electron microscopy |
| TFE | Tetrafluoroethylene |
| $T_g$ | Glass transition temperature |
| TS | Tensile strength |
| UV | Ultraviolet |
| WE | Working electrode |
| WU | Water uptake |
| XPS | X-ray photoemission spectrometry |

# Index

www.ingramcontent.com/pod-product-compliance
Lightning Source LLC
Chambersburg PA
CBHW081533220326
41598CB00036B/6420